The Effects of Hostile Environments on Coatings and Plastics

The Effects of Hostile Environments on Coatings and Plastics

David P. Garner, EDITOR
General Motors Research Laboratories

G. Allan Stahl, EDITOR
Phillips Petroleum Company

Based on a symposium
sponsored by the ACS Division
of Organic Coatings and Plastics
Chemistry at the 184th Meeting
of the American Chemical Society,
Kansas City, Missouri,
September 12–17, 1982

ACS SYMPOSIUM SERIES **229**

AMERICAN CHEMICAL SOCIETY
WASHINGTON, D.C. 1983

Library of Congress Cataloging in Publication Data

The Effects of hostile environments on coatings and plastics.
(ACS symposium series, ISSN 0097-6156; 229)

"Based on a symposium sponsored by the ACS Division of Organic Coatings and Plastics Chemistry at the 184th Meeting of the American Chemical Society, Kansas City, Missouri, September 12-17, 1982."

Bibliography: p.
Includes index.

1. Plastics—Effect of environment on—Congresses. 2. Protective coatings—Effect of environment on—Congresses. I. Garner, David P., 1950- . II. Stahl, G. Allan. III. American Chemical Society. Division of Organic Coatings and Plastics Chemistry. IV. Series.

TA455.P5E38 1983 620.1'9232 83-9230
ISBN 0-8412-0798-4

Copyright © 1983

American Chemical Society

All Rights Reserved. The appearance of the code at the bottom of the first page of each article in this volume indicates the copyright owner's consent that reprographic copies of the article may be made for personal or internal use or for the personal or internal use of specific clients. This consent is given on the condition, however, that the copier pay the stated per copy fee through the Copyright Clearance Center, Inc. for copying beyond that permitted by Sections 107 or 108 of the U.S. Copyright Law. This consent does not extend to copying or transmission by any means—graphic or electronic—for any other purpose, such as for general distribution, for advertising or promotional purposes, for creating new collective work, for resale, or for information storage and retrieval systems. The copying fee for each chapter is indicated in the code at the bottom of the first page of the chapter.

The citation of trade names and/or names of manufacturers in this publication is not to be construed as an endorsement or as approval by ACS of the commercial products or services referenced herein; nor should the mere reference herein to any drawing, specification, chemical process, or other data be regarded as a license or as a conveyance of any right or permission, to the holder, reader, or any other person or corporation, to manufacture, reproduce, use, or sell any patented invention or copyrighted work that may in any way be related thereto.

PRINTED IN THE UNITED STATES OF AMERICA

ACS Symposium Series

M. Joan Comstock, *Series Editor*

Advisory Board

David L. Allara
Robert Baker
Donald D. Dollberg
Brian M. Harney
W. Jeffrey Howe
Herbert D. Kaesz
Marvin Margoshes
Donald E. Moreland
Robert Ory
Geoffrey D. Parfitt
Theodore Provder
Charles N. Satterfield
Dennis Schuetzle
Davis L. Temple, Jr.
Charles S. Tuesday
C. Grant Willson

2-23-84 $47.95

FOREWORD

The ACS Symposium Series was founded in 1974 to provide a medium for publishing symposia quickly in book form. The format of the Series parallels that of the continuing Advances in Chemistry Series except that in order to save time the papers are not typeset but are reproduced as they are submitted by the authors in camera-ready form. Papers are reviewed under the supervision of the Editors with the assistance of the Series Advisory Board and are selected to maintain the integrity of the symposia; however, verbatim reproductions of previously published papers are not accepted. Both reviews and reports of research are acceptable since symposia may embrace both types of presentation.

CONTENTS

PREFACE

THE QUESTION ARISES: "WHAT CONSTITUTES A HOSTILE ENVIRONMENT FOR PLASTICS?" The word corrosive has long been associated with metals, and by inference, the effects of corrosives on plastics would suggest environments corrosive to metals. However, a corrosive environment for a metal is not necessarily a harmful environment for a polymer. When exposed to conditions that might corrode a metal, a polymer may not be affected or may only be nominally affected. Conversely, environments that have no effect on metals can quickly destroy a plastic or deface a coating. The hostility of an environment is a function of the material.

This volume is composed primarily of the expanded papers from the symposium, "The Effects of Hostile Environments on Plastics: A Symposium Held in Honor of Plastics Pioneer Raymond B. Seymour." Additional papers were included for better coverage of the overall topic. Each session of the symposium had a special theme: introduction and background, bulk properties, and surface characteristics.

Our goal is to describe some degenerative conditions, to detail the way polymers fail, and to give alternative types of materials that are successful in hostile environments.

We have avoided the temptation to divide the papers into convenient classes of physical (mechanical, thermal, and electrical), chemical (moisture, acids and bases, solvents, salts, and gases), atmospheric (weathering, radiation, and vacuum), and biological resistance. Although helpful, the use of these classes discourages emphasis on the basic behavior of specific materials.

In this volume, the chapters have been divided into four sections. The first serves as an overview of and introduction to the effects of hostile environments on the micro- and macrostructure of polymers. The three chapters included in this section examine heat and radiation effects, and to a lesser extent, shear and solvent effects.

Two chapters in the second section contain descriptions of specific commercially available polymers that perform under demanding conditions. Another chapter describes several "high-performance" plastics and gives general characteristics for each. The final chapter in this section is a comprehensive review of attempts to make polymers that "fight back" against attack by bioorganisms.

The seven chapters of the third section examine the effects of hostile environments on bulk properties. The environments range from acids or pickling baths to water. The materials range from polyolefins to specialty block copolymers. The final chapter in this section contains recent findings about the effect of ozone on bulk properties.

The final section on surface characteristics opens with a continuation of the effects of ozone. Two chapters report the consequences of atmospheric pollutants on surfaces. The final chapter discusses biological degradation on surfaces.

DAVID P. GARNER
General Motors Research Laboratories
Warren, MI 48090

G. ALLAN STAHL
Phillips Petroleum Company
Bartlesville, OK 74004

May 13, 1983

1

Raymond B. Seymour: Coming of Age with Polymers

W. N. TINNERMAN

University of St. Thomas, Houston, TX 77006

This Symposium on "The Effects of Hostile Environments on Plastics" is to honor a Plastics Pioneer, Raymond B. Seymour. The co-chairmen David Garner and G. A. Stahl as well as this speaker are former graduate students of Dr. Seymour.

I first met Professor Seymour in the Spring of 1972, when I was discharged from the U.S. Army. I chose to study and work under the guidance of this man. He convinced me that Polymer Chemistry was a highly rewarding discipline which took in the many areas of chemistry, physics and math. He repeatedly told me and his other students, including graduate students of other faculty members (sometimes to their displeasure), of the advantages of a background in polymer chemistry. Even the statistics indicating that more than 50% of all chemists work in a polymer related area did little to attract large numbers of students to pursue research in the area of polymer chemistry at the University of Houston. This was most unfortunate, as these non-believers missed an opportunity to study and work with a vigorous, dedicated, industrious man who has experienced success in both academic and industrial careers. He has been involved in production, research, management, achieving the top rung of the ladder as President of a chemical firm.

I would like in the brief time allotted to me to give you a thumbnail sketch of the person we are honoring. It would be impossible to go into detail, particularly as I am only personally familiar with Dr. Seymour's achievements in the last 10 years. Therefore, I have taken the liberty to extract information from letters of nomination as well as seconding letters for the various regional and national awards Dr. Seymour has received along with his personal vitae.

Professor Seymour was born on 26 July 1912 at New England Baptist Hospital in Boston, Massachusetts. He was the second of four children. He attended grade schools in Lexington,

0097–6156/83/0229–0001$06.00/0
© 1983 American Chemical Society

Figure 1. Ray Seymour in 1932 during his first college years in Dover, NH.

Massachusetts; New Fields, New Hampshire and Dover, New Hampshire. He graduated from Dover High School in the Spring of 1929. He was selected as an Edison schollar in 1929, and was personally examined by Thomas Edison. He entered the University of New Hampshire in the Fall of 1929, where he majored in Chemical Engineering. At the time, he didn't realize the abbreviation for Chemical Engineering was Ch.E. rather than C.E. and he ended up with a class of civil engineers in his first quarter, which was corrected in the second quarter. He received an education in both chemistry and chemical engineering and has lead a dual life as both a chemist and chemical engineer. While at the University of New Hampshire, Raymond Seymour was under the guidance of Dr. Harold A. Iddles, which Dr. Seymour states was an outstanding teacher whose qualities he has tried to emulate. Dr. Seymour must have succeeded, because in 1976 he was awarded the Excellence in Teaching Award by the Chemical Manufacturers Association. This prestigeous award had also been presented previously to Dr. Iddles.

He received his B.S. in Chemical Engineering with high honor in June 1933, whereupon he was accepted as a graduate assistant at the Graduate School of University of New Hampshire.

He completed his Master of Science in Chemistry from the University of New Hampshire in 1935. He entered the University of Iowa in the Fall of 1935. As a member of Alpha Chi Sigma, which he joined at University of New Hampshire, he was able to live in the Alpha Theta Fraternity House at Iowa City which was next to a dormitory for girl students. Here he met Francis Horan who he married in September 1936. Francis and he will be celebrating their 46th anniversary this month. The first of their four children was born 1 month before Dr. Seymour received his Ph.D. degree in Chemistry in the late summer of 1937.

The industrial career of Dr. Seymour was now underway, as he had already interviewed for and accepted a job in April 1937 with Universal Oil Products, paying $250 a month.

He, Francis and their 1-month old son, David, drove from Iowa City, Iowa to Riverside, Illinois in a Model A where he was to begin his job at the U.O.P. laboratory. He was met by Dr. Gus Eggloff and told there wasn't any job because the chemical engineers and architects were on strike. As a professional chemist and President of the American Institute of Chemists, Dr. Eggloff suggested that Dr. Seymour look in the want ads of the Chemical Engineering News. Continuing East, Dr. Seymour drove to Akron, Ohio where he interviewed with Dr. Loren Sebrell at Goodyear Tire & Rubber Co. on Monday morning and was at work in the laboratory that afternoon. But because of the depression, his salary was only $210 per month.

Figure 2. The Seymours—Frances, David, and Ray—in 1940 on an outing in Allentown, PA.

At Goodyear, he was a research chemist, working in the area of rubber accelerators, but he turned his attention to the production of copolymers of vinyl chloride and synthetic rubber. After one year, he and all the other employees were informed that their salaries had been reduced by 10%. After voicing his complaint, Dr. Seymour resigned and took a position as Chief Chemist for Atlas Mineral Products in Mertztown, Pennsylvania at the salary of $225 per month. He was their only chemist, developing a product line of protective coatings based on PVC, and pioneered the development of polymer concrete.

One year at Atlas Mineral Products, and his salary was increased to $250 per month. (Thus, it had taken him 4 years to overcome the effect of the strike at U.O.P.) But along with the raise came the request to spend less time on product development and more time in management. He resigned and took a position as Research Group Leader at Monsanto Co. in Dayton, Ohio in 1941. At Monsanto, Dr. Seymour worked with Dr. George Ham, Dr. Harry Szmant (Chairman of Chemistry of University of Detroit), Dr. Ray Meyers (Professor of Kent State University) among others.

Figure 3. Ray Seymour on the lecture circuit in 1945 in Chattanooga, TN— and he is still talking.

Figure 4. The young executive in 1948 in New Brunswick, NJ.

While at Monsanto, he was confronted with the philosophy (which later changed) that chemists at Monsanto were not permitted to publish papers. He resigned in 1945 to accept the position as Director of Research at the University of Chattanooga, at Chattanooga, Tennessee. In January 1949, he returned to Atlas Mineral Products, this time as President.

He found that Atlas' products were those he had developed in 1939, but the sons of the founder did not know the formulas or how to manufacture the material. During his 5 years at Atlas, Dr. Seymour was involved in the development of plastic pipe, formed and welded PVC structures, polyurethane coatings and many new concepts in corrosion resistant equipment. He left Atlas in 1955 to become President of Loven Chemical near Los Angeles, California. While in California, he was an Adjunct Professor at Los Angeles Trade Tech. College. In 1960, having been the President of two chemical companies during the last 10 years, he resigned and once again entered his real love, the field of academia. He accepted the position of Chairman of the Department of Chemistry at Sul Ross University in Alpine, Texas in 1960.

He was a one man department. He taught 40 semester hours a week. Class began at 6 a.m. and some ended at 10 p.m. His enthusiasm for science and education was contagious. The students seemed to enjoy it, as they were very much above average. The Raymond B. Seymour Award is given annually to Sul Ross' outstanding chemistry student. In 1964 he accepted a position as Associate Chairman of the Department of Chemistry and Associate Chairman of University Research at the University of Houston in Houston, Texas.

Two of his students at Sul Ross followed him to Houston and became his first graduate students. They are Dr. Peter Tsang, presently at Bendix Corporation, and Dr. Jose Sosa of Vulcan Chemical Company of Wichita, Kansas. In 1976 he was forced to retire from the University of Houston because of his age. He was made a Professor Emeritus, but not willing to quit, he accepted a position as Distinguished Professor in the Coatings and Polymer Science Department at the University of Southern Mississippi. He and Mrs. Seymour now have a home adjacent to the golf course in Hattiesburg.

Dr. Seymour's personal qualities are not confined to the formal classroom, it is found in the research laboratory, in the industrial laboratory, in the continuing education of professional chemists and secondary chemistry teachers, in presenting chemical education through radio and television programs, and in his participation in educational activities of the ACS and other professional organizations (such as American Institute of Chemists, American Institute of Chemical Engineers since 1945,

Figure 5. Ray Seymour back at college in 1970; this time in Alpine, TX.

Charter member of Society of Plastic Engineers, Lifetime member of American Association for the Advancement of Science, Society of the Plastics Industry).

Just within the ACS (a 50-year member), he has served in a variety of capacities:

Member since 1932
Secretary Dayton Section 1943-45
Chairman elect Dayton Section 1945
Councilor Chattanooga Section 1945-48
Member Professional Training Committee 1946-48
Member Executive Committee Organic Coating and Plastics Committee 1948, 1975-6
Chairman Elect Lehigh Valley Section 1955
Chairman Permian Basin Section 1964
Councilor Southeastern Texas Section 1976
Member Executive Committee Polymer Division 1974
Member Public Relations Committee 1973-76
Tour Speaker 1945, 1948, 1968, 1972, 1981, 1982, 1983
Abstractor Chemical Abstracts
Radio Commentator for Men and Molecules
Participant for ACS TV shows
Chairman Symposium at Mexico City, 1975; San Francisco, 1976; New Orleans, 1977; Anaheim, 1977
Invited speaker at most ACS meetings
Chairman National Awards Committee, 1977
Member ACS Advisory Committee, Advances in Chemistry Series 1978

Chairman ACS Polymer Chemistry Syllabus Committee 1980
Co-Chairman ACS Polymer Education Resources Committee 1981
Chairman Education Committee, Division of Polymer Science 1982
Member Alpha Chi Sigma since 1932

Dr. Seymour has been awarded 45 patents and is the author or co-author of 24 books and two audio courses. He has authored more than 1000 research and technical articles and was recognized by Polymer News in 1980 as the "world's best known polymer chemist". He has served as a consultant for some 12 or more companies and has served as a visiting professor of polymer science in Bangladesh, India, Taiwan, Yugoslavia, Czechoslovakia, Trinidad, USSR and Australia.

Dr. Seymour has received a large number of honors and awards which include the following:

Western Plastics Award 1960
University of Houston Teaching Excellence Award 1975
Plastic Pioneer Award 1975
Chemical Manufacturers Association Catalyst Award 1976
American Institute of Chemists Honor Scroll in Louisiana 1980
Inducted into the Western Plastics Hall of Fame 1981
Southern Chemists Award 1981
Society of Plastics Engineers International Education Award 1982.

Figure 6. Frances and the Chief, familiar faces at ACS meetings, in 1982 in Kansas City, MO.

Raymond B. Seymour is known to many friends and associates as Ray, a Plastics Pioneer, as Dad to his four children and by a variety of grandfatherly terms by his 10 grandchildren, but to his former students; including, Peter Tsang, Jose Sosa, Don Owen, Clem LaSada, Hubert A. Wood, Murray Clark, G. Allan Stahl, David P. Garner, Earl Johnson and me, he is "The Chief".

RECEIVED June 7, 1983

INTRODUCTION AND OVERVIEW

2

Degradation of Polymers in Hostile Environments

HERMAN F. MARK

Polytechnic Institute of New York, Brooklyn, NY 11201

The influences of "hostile environment" on polymers - heat, irradiation, mechanical and electrical abuse, the action of aggressive chemicals - may be generally classified in: a) changes of the chemical character and microstructure of the individual macromolecules, and b) changes in the supermolecular structure and texture of the system.

a) These changes include: decrease of the molecular weight by chain scission which causes a reduction of tensile strength, modulus and toughness, introduction of reactive (polar) groups which cause changes in compatibility, electrical and optical behavior, introduction of light-absorbing groups which cause discoloration, and internal cyclization of the chains which may result in hardening and decrease in toughness.

b) These changes include: crosslinking between hitherto independent macromolecules; a moderate extent may have favorable consequences but too much of it causes reduced impact strength and brittleness, additional crystallization; here again a certain degree of this change may be advantageous whereas too much may cause opacity, haze and crack formation, and release of gaseous fragments causing bubble formation and reducing strength.

A few examples for particularly drastic influences of hostile environment on polymeric systems will be discussed.

It is a pleasure and an honor for me to be contributing to the proceeding of a symposium held in honor of my good friend,

0097-6156/83/0229-0011$06.00/0
© 1983 American Chemical Society

Ray Seymour. We first met at Gibson Island some forty years ago at a Gordon Research Conference. Since then, Ray has had an illustrious and eventful career; he has been an understanding experimental chemist, a very successful manager and advisor to chemists, and now an excellent educator of chemists. I think that anyone who has followed his career is aware of his numerous contributions to chemistry and we all hope that in his current position at the University of Southern Mississippi he will continue to promote our science as well as make the world a better place to live. Ray is a true gentleman as well as a shining example for a fine scientist.

Discussion

I have been asked to give an overview of the effects of hostile environment on plastics. Emphasizing not the degradation or destruction of the bulk polymeric material and its properties, but rather what is happening to individual macromolecules in hostile environments, and also to larger assemblies of them. The hostile environments to be discussed are heat, radiation, mechanical stresses and the presence of oxygen in combination with heat and radiation.

Heat Degradation. Heat is molecular motion and for solid bodies may be represented by waves which traverse the system in directions. Waves can be very complex and there can be interference phenomena and fluctuations associated with them. A simple example is the white caps on a lake. White caps are the constructive interference of water waves. Each individual wave would not have the energy to form a white cap, but a combination of waves does yield them. Thus, it is the concentration of energy at a given point that is important. As it is in water, so can it be in a macromolecule. It is not the motion due to the average temperature of the polymeric material, but the fluctuations that are critical. If the fluctuations are frequent and there is a concentration of energy through constructive interference, then there is a concentration of energy at a point along the chain that is higher than the strength of the chain can support.

In a long flexible chain molecule there are various segmental motions. Eventually at one point there will be such a concentration of energy that the chain breaks. In the simplest case, that of polyethylene (PE) with only C-C and C-H bonds, the disassociation energy for these bonds is about 80 and 90 kcal/mol, respectively. So when the energy exceeds the amount, the chain breaks, the two sigma bonding electrons are separated and two lone electrons (free radicals) are formed.

A very probable reaction would be for the radicals to recombine. They are already in a bonding situation. Since that is occurring in solid system (or concentrated solution), there is a possibility of a cage effect. That is, after separation, the free radicals are held in a relatively confined area and as a result recombination is likely. This is particularly true if the chains are rigid and typically rigid chains containing cyclic aromatic units in the chain have high heat stability. Another example of chains being held in very rigid position is in diamond. A carbon-carbon bond in diamond is no stronger than one in PE or any other organic compound. But in a diamond with its three dimensional structure, when a bond breaks there is little that the radicals can do but recombine. So, even a very weak, sensitive bond may be protected to some degree by a cage effect if it is in a sufficiently rigid environment.

On the other hand, if the chains are flexible like PE and the temperature is high with the segments moving around quite rapidly, then the two electrons on the chain end may become so separated that it is unlikely that they would recombine. Two free radicals are then dormant. A radical may pick up a hydrogen from an adjacent chain and the radical is transferred. However, even though the chains are broken, the amount of material present remains the same. That is, there is a reduction of molecular weight (MW) but not of weight. The number of chains go up and the MW of the chains goes down. There is a reduction in the properties of the material, but again the total weight of the sample does not go down.

What happens to the free radicals? If they are on a flexible molecule like PE which is moving around quite rapidly, then there is a high probability that it may pick off a hydrogen of its own chain five to seven carbon atoms back along the chain. So then there is a transfer of the radical to a position away from the chain end and a termination of activity at a chain end. The free radical may then pick up a monomer and then continue growing from the new reactive position. The net effect is that a short branch has been formed. In principal, this is similar to the well known short chain branching of PE when produced by the high temperature-high pressure method. In this case, many short branches are formed. There is also a probability that the free radical may attack an atom farther removed along the same chain and then cyclization may occur.

So, in the simplest case we may expect degradation, MW loss, branching and cyclization.

Now things get more complicated as soon as there are substituents on the chain. Again, a bond may be broken this time between a CH_2 unit and a CHR unit where R is the substituent. Two different types of radicals are then formed. The CH_2 radical end group is a very reactive radical and soon will pick up a hydrogen. The CHR radical is more stable and may even be very stable depending on the group R. If it is relatively stable, then it may disassociate into a monomer and initiate a stepwise chain depolymerization. There are several cases where this is very pronounced, particularly if R is very stabilizing as in the case of alpha-methylstyrene. The degradation in this case is of a different nature. The monomer is eliminated and therefore the weight of the sample goes down while the MW of the individual molecule remains approximately the same. Taking two measurements, namely the weight of the material and the MW of the chains, allows us to distinguish between the two degradation mechanisms of chain scission and depolymerization in terms of relative frequency of the two mechanisms. For example, based on this approach, we know that poly-(methyl methacrylate) (PMMA) can be degraded at high temperatures to give a good yield of MMA monomer.

A third possibility is that the substituent reacts with a neighboring hydrogen and RH is released. Whenever the bond strengths of RH are larger than the bond strengths of the chain and the temperature is high enough, this will be an additional means of decomposition. All three types of decomposition discussed thus far apply to a single chain, and under given conditions, may occur simultaneously.

If we consider that the chain is one of many, then when a free radical is formed because of chain scission somewhere in the system a hydrogen is abstracted with the result that the radical may be transferred to another chain. If a chain has more than one free radical and if the chains are in rapid motion, then the radicals may be shifted along the chain and can possibly wind up next to one another. The radicals may then combine to form a double bond.

The main features of degradation for individual chains due to high temperature then are: chain scission, depolymerization, branching, self branching, self cyclization, removal of a substituent, and formation of double bonds.

Radiation Degradation. What if irradiation of the molecule occurs, and what is the difference between heating a material and irradiating it? Heat essentially works on the vibrations of atoms; it is uniform throughout the system. Radiation works on the electrons and only a few sites will absorb energy. The energy in the form of photons does not go directly into the vibrational modes of the molecules, but instead into the excitation of certain of its electrons. All other sites remain uneffected. From the theory of electronic excitation we know that if somewhere in a molecule an electron is put into an excited state, there are three ways to get rid of the excess energy. One is fluorescence, one is phosphorescence, and the last and most important for this discussion is the cascading down of the electronic energy into vibrational and rotational energy; energy embodies in these modes of atoms can cause the breaking of a bond.

Radiation covers a wide range of frequencies from radio to gamma rays. The frequencies of radiation that are most harmful to polymeric systems are those from the blue part of the visible spectrum and the near ultraviolet. The longer wavelengths are not energetic enough to harm molecules and most of the other potentially harmful high frequency rays are screened out by our atmosphere.

However, at 300 nm and above into much radiation penetrates the atmosphere and can act on the material. Here groups such as

C=O, C=C, and N=C begin to absorb and electrons are lifted into higher energy levels. When these electrons decay to lower energy states, energy is released in the form of vibrational modes which may cause a bond to break and create free radicals. After the radicals are formed they will again react as has been previously described. The one additional danger is that the excited state may be long lived, which causes a degradation after a long period of time.

Mechanical Degradation. A third danger is mechanical degradation. Any kind of mechanical action which changes substantially the shape or size of a polymeric sample has the potential to break single bonds. The question is how fast do they form and how fast do they disappear by picking up hydrogen or by cyclization. Anything that cuts, grinds, or shears has the capability of generating free radicals.

The mastication of rubber is a good example. The purpose of mastication is chain scission and the reduction of the average MW of the rubber.

Another example in a different system would be the following: Suppose we have a semicrystalline material and there are chains which connect two crystalline domains of that material with each other. There is no reason why the lengths of those chains which are in the amorphous area between the crystalline site should be the same. There will always be a distribution of chain lengths. Then if one keeps one crystal constant and moves the other as soon as the stress is greater than the bond strength of the shortest chain, a rupture of this chain will occur. As the distance between the crystals increases, more and more chains will break. Eventually, a crack will form which may then propagate by additional processes.

The Rule of Oxygen in Polymer Degradation. A last point to consider is that in most cases heat or radiation does not act alone, but acts in conjunction with oxygen. The additional problem is that oxygen is a very reactive molecule and if a free radical forms in its presence then it can combine immediately with it to form a different radical. This radical may then abstract a hydrogen and form a hydroperoxide. The hydroperoxide can then decompose into two radicals and so from the initial two radicals a total of six possible radicals may form. This explains the danger in chain scission in the presence of oxygen leading to a chain reaction.

Summary

So, in essence, all four types of degradation proceed through free radical reactions. Depending on the conditions, they can be relatively harmless when a cage effect is possible or there can be a whole sequence of secondary and tertiary reactions which lead to polymer degradation.

RECEIVED April 11, 1983

3

Effects of Radiation Environments on Plastics

MALCOLM DOLE

Los Gatos Meadows-M101, Los Gatos, CA 95030

The effects of radiations in the environment on plastics are discussed from the standpoint of (1) major chemical reactions produced by ionizing radiations, electron beams or cobalt-60 gamma rays; (2) effects of the radiation on the physical properties of plastics; (3) relative sensitivity of different plastics to the environment; and (4) additives that can be incorporated into plastics to increase their resistance to radiations.

Radiation treatment of plastics as in the use of radiation for sterilization of medical supplies, may be more protective of the environment than sterilization by heat. For example, in the curing of rubber sheet ten times as many Btu's/kg of steam are required than by using electron beams. The pollution of the atmosphere from burning coal or petroleum may be more serious than pollution arising from the production of electron beams (1). However, one usually thinks of the dangers of radiation from nuclear reactors. In fact, it was the discussion of problems arising from the effect of intense radiation on the wiring insulation within the containment vessels of nuclear reactors in classified documents in 1944 and 1945 that led the author during the academic year 1947-8 to begin his studies of the effect of radiations on the plastic polyethylene (PE). In this very early work we found that when PE was exposed to the radiation of the heavy water pile of the Argonne National Laboratory, the PE was oxidized to the extent of production of carbonyl groups if the irradiations occurred in air (2). We also showed that unsaturation was produced due to hydrogen evolution by observing the increase in weight of the irradiated PE films when exposed to bromine vapor and also the increase in concentration of the trans-vinylene infrared absorption band at 966 cm^{-1}. Most surprising to us was the creation by the irradiation of co-valent C-C bonds (crosslinks) between the long $-CH_2-$ chains in the solid state at room temperature. The extensive work done by Charlesby and co-workers has been excellently

0097-6156/83/0229-0017$06.00/0
© 1983 American Chemical Society

described in Charlesby's treatise published in 1960 (3). Although high energy radiations can produce many deleterious effects, there are some beneficial aspects resulting chiefly from production of crosslinks that have led to industrial applications of great importance.

The radiation environments with which we shall be concerned are the intense radiation fields within nuclear reactor containment vessels, the radiation produced by ^{60}Co gamma rays, the environment in the neighborhood of high energy electron beams which may contain ozone and cosmic rays although we shall not discuss the latter because the radiation intensities resulting from cosmic rays are very weak, especially at low altitudes and within buildings. Even commercial airline personnel flying many hours at high altitudes do not seem to be affected by cosmic rays.

Major Chemical Effects

We begin a listing of chemical effects with a summary of the gases produced by irradiating plastics. Because most plastics contain hydrogen, we find that hydrogen is evolved on exposing the plastics to high energy radiations. Table I compares the hydrogen yields for a number of plastics irradiated in a vacuum at room temperature; the yields expressed in units of G-values, i.e., $G(H_2)$ equals the molecules of hydrogen evolved per 100 e.v. of energy absorbed (4).

Table I. G-values for Gases Produced by γ-Irradiation.

$G(H_2)$		$G(H_2)$		$G(HC\ell)$	
LDPE	3.9	PPOx	1.12	PVC	13
PP	2.7	PMMA	0.1		
PS	0.022 - 0.026	Nylon 6-6	0.36		
POM	1.7	PET	0.02		
PVAc	0.64				

From Table I it is clear that polystyrene (PS) and polyethylene terephthalate (PET) are much more resistant to γ-rays than polypropylene (PP), low density PE (LDPE), poly(oxymethylene) (POM), poly(vinyl acetate) (PVAc), poly(propylene oxide) (PPOx), and somewhat more resistant than poly(methyl methacrylate) (PMMA) or 6-6 nylon. A very commonly used plastic, poly(vinyl chloride) (PVC), is perhaps the least resistant of all the plastics giving off hydrogen chloride with quite a high G-value when irradiated.

To produce hydrogen from PE by irradiation, it is necessary that the C-H bond be broken. Inasmuch as the chemical bond energy of the C-C covalent bond is less than that of the C-H bond, at first sight it might be expected that the C-C bond in PE would undergo scission more frequently than the C-H bond. This is not the case, however, as the G-value for scissions G(S) is quite low, about 0.1 to 0.2 in linear low density polyethylene, LLDPE, as we have found recently (5). Also in polybutadiene no chain scission on irradiation can be detected (6, 7). When the polymer chain contains carbon atoms connected to four carbon atoms as in poly(isobutylene) (PIB), or to three as in PP, then considerable chain scissions are produced by the irradiation, G(S) equal to ~0.3 for PP (8) and ~3 for PIB (4). As indicated by the data of Table I, it could be expected that G(S) for PS would be quite low and this turns out to be the case; G(S) equal to 0.01 to 0.02 (4) despite the fact that in PS every other carbon atom along the chain is connected to three carbon atoms. PMMA undergoes considerable chain scission, G(S) equal to 3.5 for irradiation in vacuo at room temperature. An interesting aspect of the radiation chemistry of PMMA is that the isotactic and syndiotactic forms of PMMA are also converted to the heterotactic structure (8) on irradiation.

On breaking the C-H bond in PE an alkyl free radical of the structure $-CH_2\dot{C}HCH_2-$ results. When two free radicals on neighboring chains combine a C-C covalent bond crosslinking the two chains is formed. The G-value for crosslinks, G(X), is of the order of magnitude of 2.5 in quenched LLDPE of density equal to 0.9175 at 25° (5). The evidence is that the crosslinks occur primarily in the amorphous phase of the PE as well as in the amorphous surface layers of single crystals of PE. In HDPE G(X) is about 1.5 depending on the dose (9). The greater the concentration of double bonds in the polymer, the higher G(X) is. For example, in poly(butadiene) (PB) G(X) is 5.7 for irradiations at 25° in vacuo (6).

The liberation of hydrogen from PE, $G(H_2)$ about 3.7 (9), also produces trans-vinylene unsaturation. In addition to the formation of this unsaturation the radiation also causes its disappearance and that of other unsaturated groups. In fact in LHDPE of the Marlex type the G-value for the decay of the vinyl end groups, G(-Vi) is equal to 9.6 (10) at zero dose for irradiation in vacuo at 35°. As the concentration of the vinyl groups is about 1000 times less than that of the CH_2 groups, one would have expected a far smaller G(-Vi) than $G(H_2)$. But because G(-Vi) and $G(H_2)$ are the same order of magnitude there must be considerable energy transfer along the PE chains to the vinyl groups during its irradiation.

From the standpoint of the effect of radiation in the environment on plastics, perhaps the most important reaction is that of oxidation. By measuring the increase in the infrared absorption band of the $>C=O$ group at 1725 cm^{-1}, Dole and his co-worker, Rose, discovered in the very earliest studies (2) on PE that considerable oxidation occurred for irradiations of their thin PE films in air. Matsuo and Dole (11) measured the G-value for carbonyl formation, G(C=0), and found 5.0 for LHDPE. Dole (12) estimated a value of 6.3 from studies of Giberson (13) on a LDPE film. Not only does the oxidation occur during the irradiation, but it may continue for weeks after the irradiation if the allyl free radicals of structure $-CH_2-\dot{C}HCH=CH_2$ or $-CH_2\dot{C}HCH=CHCH_2-$ are not annealed out by heating to 100° or higher after the irradiation. Thus Williams working in the author's laboratory (14) showed that even after sixty days of exposure to air at room temperature the optical density of the carbonyl band at 1725 cm^{-1} was still increasing. Ohnishi, et al., (15) have suggested mechanisms for two chain reactions by which the oxidation can occur. One of the mechanisms leads to chain scission with the final products, $-CH_2CHO$ and $-CH_2\dot{C}HCHO$. The latter free radical then adds oxygen and the process can continue. The other mechanism involves a hydrogen-atom shift to form a hydroperoxide group, which then rearranges to yield as the final product

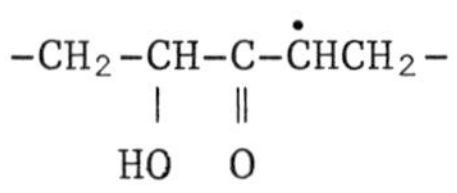

which can also add oxygen and continue the process as a chain reaction. Zhdanov, et al., (33) quote a G-value of 0.9 for the formation of peroxide macroradicals in the irradiation of poly(methyl methacrylate) at 273K.

It should be emphasized that oxygen gas dissolves only in the amorphous regions of polymers, and can, therefore, oxidize the material only within the amorphous phase or on the surface. Also the extent of oxidation per unit of radiation dose is very dependent on the dose rate. The lower the dose rate the more time the oxygen has to diffuse into the plastic per unit of dose so that the net amount of oxidation for the same dose is much greater at low rates.

In this connection it is interesting to note that Dvornik and co-workers (17) have recently proposed a low-level chemical dosimeter made of 10 vol. % chlorobenzene, 10 vol. % ethanol in 2,2,4-trimethylpentane using a spectrophotometric method for $C\ell^-$ determination. This should be of considerable use in measuring the intensity of ionizing radiation when of a low-level.

Effects of Radiation on Physical Properties

As can be imagined the radiation produced crosslinks and chain scissions act in opposite directions with respect to changing the physical properties of plastics. Crosslinks transform molten polyethylene from a liquid into an elastomer. The elastomeric properties thus produced make possible the so-called "memory effect" in which the irradiated PE, for example, when stretched at temperatures above its melting point and then cooled back to room temperature while still stretched will subsequently "remember" its original unstretched dimensions when heated above its melting point. It will then spontaneously shrink back to its original dimensions and shape. The beautiful thing about this process is that the crosslinks can be formed in the solid state at room temperature after the plastic has been molded into any desired shape. The radiation crosslinking does not affect the shape of the object being irradiated.

In our first studies on PE (2) it was found that the irradiations of the heavy water pile completely removed the ability of strips of PE film to be cold drawn. In other words, the stress-strain curves were straight up to break. The marked change in this physical property was interpreted in terms of the formation of crosslinks. This explanation was confirmed by Charlesby (16) who showed that the irradiation of PE made part of the material insoluble because of the formation of three dimensional network structures. The radiation induced crosslinking in plastics has the practical effect of increasing the stability of electrical insulation at higher temperatures and the radiation treatment of such materials is of considerable industrial importance. As an example, Campbell (18) states that aging at high temperatures in gamma-ray radiation produced a considerably longer lifetime in a polyimide, a poly(vinyl formal) and a polysiloxane than thermal aging alone when compared at the same temperature. Wilski and co-workers (19) have recently measured the ultimate tensile strength of high impact polystyrene as a function of the ^{60}CO gamma ray dose in the absence of oxygen and found it to increase linearly with dose up to 3 MGy (300 Mrad). During the irradiation crosslinks produced in the polybutadiene phase reduced the elongation at break to half of its initial value after a dose of 1.9 MGy and the impact strength to half of its initial value after a dose of 2.3 MGy. A very long ten year exposure in air of the high impact PS to a low gamma ray dose rate of 3.72 mGy/s (0.00134 Mrad/hr) reduced the ultimate tensile strength instead of increasing it to half of its initial value after a total dose of 0.9 MGy (90 Mrad). The reduction instead of an increase in the tensile strength resulted, of course, from the oxidation that occurred. Storage of the high impact PS in the dark for fourteen years did not change its mechanical properties.

Although radiation may raise the softening temperature because of the crosslinks produced, the thermodynamic melting point is hardly changed in PE as demonstrated by Dole and Howard (20) by means of accurate specific heat measurements. Ultrahigh molecular weight polyethylene, UHMWPE, is used as implantable prosthetic devices such as total joint replacements because of its mechanical properties, resistance to chemicals and to abrasion. Irradiation enhances its creep resistance (21).

Resistance of Polymers to Radiation

Poly (vinyl chloride), PVC, is quite sensitive to radiations; the G-value for evolution of hydrogen chloride has been determined to be 13 at 30° and 23 at 70° (22). Shin and Paik (23) found that PVC plasticized with dioctyl phthalate was more resistant to radiation than rigid PVC. Campbell (18) in his table comparing the radiation stability of fifty-five plastics and elastomers lists polytetrafluoroethylene (PTFE) as the least resistant. In 1969 Hedvig (24) showed that films of PTFE of 0.04 mm thickness had their mechanical strength reduced from 269 to 136 kg cm^{-2} and their elongation at break reduced from 129 percent to 10 percent by an irradiation to a dose of 10.4 Mrad in air. Wall and Florin (25) compared the embrittlement of PTFE on irradiation in air with that on irradiation in a vacuum or an inert atmosphere and found that in the latter case the critical absorbed dose to failure was 100 times greater than with oxygen present. El-Sayed, et al., in studying the radiation grafting of acrylic acid onto PTFE (26) and onto the copolymer polytetrafluorethylene-hexafluoropropylene (FEP) (27) found that more grafting was obtained on FEP than on PTFE, that the FEP film had a considerably higher radiation resistance than PTFE films and that FEP can be irradiated without extensive degradation. The Hitachi Cable Co., of Japan has a patent (28) on a mixture of 100 parts PE and 5 parts of an ethylene-1, 4-hexadienepropylene copolymer irradiated with electron beams to 15 Mrad dose after which it had an insoluble gel fraction of 0.75 and 19 percent heat distortion compared to a gel fraction of 0.60 and a 28 percent heat distortion after an electron beam dose of 20 Mrad in the case of a pure PE sample (28). The method of measuring the heat distortion was not described in the abstract.

Inorganic fillers can sometimes improve the radiation resistance of polymers. Kuznetsov and co-workers (29) investigated 30 PE-rubber compositions filled with carbon black and found that the gasoline resistance increased 20 to 40 percent on increasing the dose to 7 Mrad and that the thermal stability increased up to 21 Mrad. In the case of ultra-oriented high density PE, Tincer, et al., (30) found that drawing reduced the oxidation of the sample for irradiation in air at 80° (probably because of a reduced oxygen solubility in the drawn sample). Adding carbon black increased the modulus initially, but after doses in the range of 40 to 60 Mrad, the modulus decreased rapidly with dose. Carbon black did not affect the gel content, but acted antagonistically to antioxidants in that oxidation was higher in samples containing both carbon black and the antioxidant. Clough and Gillen (31, 32) studied several types of polymeric insulation materials with respect to their degradation at low dose rates of Co-60 γ-rays. The experiments were performed on PE insulation and carbon black filled PVC jacketing similar to that used in nuclear applications. They point out that from tests of insulation stability at high dose rates conclusions regarding the long term stability of the insulation at low dose rates should not be drawn because of the greater degradation per unit of dose at the low dose rates as mentioned above.

Conclusions

It is evident from the above that the effect of high energy radiations such as Co-60 gamma rays or high speed electrons on plastics is a very complicated subject with the predicted effects depending greatly on the composition of the plastic. Nevertheless, it is clear that linear chains containing phenyl group side chains are the most resistant plastics to ionizing radiation. Polytetrafluorethylene and polymers with aliphatic side chains such as polypropylene and polyisobutylene are the least resistant to radiation. Campbell (18) ranks polyformaldehyde as one of the polymers most susceptible to radiation damage, but polyethyleneterephthalate as quite stable, although it does undergo embrittlement. The relatively high stability of the latter no doubt arises from the benzene ring in the main chain.

We conclude that with many options available regarding the composition of plastics, it is possible to design the environment of nuclear reactors so that a rather long life of the component parts can be realized. The goal is to have equipment that will last over the lifetime of the plant, projected to be forty years, and that will withstand an estimated integrated dose of $2x10^5$ Gy (20 Mrad) (18).

Literature Cited

1. Mainati, R.L.; Schmidt, C.K.; Peters, R.J. Int. Conf. on Radiation Processing for Plastics and Rubber. Brighton, U.K. 1981, 14.1.
2. Dole, M. Report of Symposium IV. "Chemistry and Physics of Radiation Dosimetry." Technical Commend, Army Chemical Center, Maryland, 1950, p. 120.
3. Charlesby, A. "Atomic Radiation and Polymers"; Pergamon Press, New York, 1960.
4. Dole, M., Ed. "The Radiation Chemistry of Macromolecules" Academic: New York, 1973 Vol. II.
5. Basheer, R.; Dole, M. Submitted for publication.
6. Basheer, R.; Dole, M. Macromol. Chem. 1982, 183, 2141.
7. Batsberg, W.; Kramer, O. J. Chem. Phys. 1981, 74, 6507.
8. Thompson, E.V. Polym. Lett. 1965, 3, 675.
9. Kang, H.Y.; Saito, O.; Dole, M.: J. Am. Chem. Soc. 1967, 89, 1980.
10. Dole, M.; Milner, D.C.; Williams, T.F.; J. Am. Chem. Soc. 1958, 60, 1580.
11. Matsuo, H.; Dole, M. J. Phys. Chem. 1959, 63, 837.

12. Dole, M. "Advances in Radiation Chemistry"; Burton, M.; Magee, J.L., Eds. John Wiley, New York, 1974, Vol. 4, p. 307.
13. Giberson, R.C. J. Phys. Chem. 1962, 66, 463.
14. Ref. (12) p. 379.
15. Ohnishi, S.-I; Sagimoto, S.-I.; Nitta, I. J. Polymer Sci. A. 1963, 1, 605.
16. Charlesby, A. Proc. Roy Soc. (London) 1952, A215, 187.
17. Vekic, B.; Razem, D; Dvornik, I. Abstracts 5th Tihany Symposium on Radiation Chemistry 1982, p. 178.
18. Campbell, F.J. Proc. 3rd Int. Meeting on Radiation Processing 1980, 103.
19. Wilski, H.; Gaube, E.; Rosinger, S. Colloid & Polymer Sci. 1982, 260, 559.
20. Dole, M.; Howard, W.H. J. Phys. Chem. 1957, 61, 137.
21. Nusbaum, H.J.; Rose, R.M. J. Biomed. Materials Res. 1979, 13, 557.
22. Salovey, R. "The Radiation Chemistry of Macromolecules." Dole, M., Ed. Academic: New York 1973, Vol. II, p. 38.
23. Shin, C.S.; Paik, Y.H. Chungi Hakhoe Chi 1981, 30, 639.
24. Hedvig, P. J. Polym. Sci. Part A-1, 1969, 7, 1145.
25. Wall, L.A.; Florin, R.E. J. Appl. Polym. Sci. 1959, 2, 251.
26. El-Sayed; Hegazy, A.; Ishigaki, I.; Okamoto, J. J. Appl. Polym. Sci. 1981, 26, 3117.
27. El-Sayed; Hegazy, A.; Ishigaki, I.; Dessouri, A.M.; Rabie, A.; Okamoto, J. J. Appl. Polym. Sci. 1982, 27, 535.
28. Japanese patent 81 159 237, 1981; Chem. Abstr. 1982, 96, 40.
29. Belyakova, A.M.; Gusev, V.I.; Pavlii, V.G.; Kuznetsov, E.V. Komoz. Polim. Mater. 1981, 9, 44.
30. Tincer, T.; Cimen, F.; Cimen, I.; Akay, C. Abstracts Int. Conf. on Radiation Processing for Plastics and Rubber, Brighton, U.K. 1981, 33, 1.
31. Clough, R.L.; Gillen, K.T. Radiat. Phys. Chem. 1981, 18, 661.
32. Gillen, K.T.; Clough, R.L., ibid. 1981, 18, 679.
33. Zhdanov, G.S.; Millinchuk, V.K.; Khamidova, L.G. Abstracts 5th Tihany Symposium on Radiation Chemistry 1982, p. 195.

RECEIVED February 4, 1983

4

Thermal Analysis of Metal-Containing Polymers: Generalizations

CHARLES E. CARRAHER, JR.

Wright State University, Department of Chemistry, Dayton, OH 45435

The topic of thermal characterizations of inorganic and organometallic polymers was recently reviewed by the author in a comprehensive review (1). The review emphasized particular findings and included a review of phosphorus-containing polymers and catenation polymers.
The current chapter will emphasize general trends with respect to both instrumentation and results for metal-containing polymers with the realization that exceptions exist for most generalizations.

The usual thermal techniques of thermal gravimetric analysis (TG), differential thermal analysis (DT), differential scanning calorimetry (DSC), thermal mechanical analysis (TM), torsional braid analysis (TB) and pyrolysis gas chromatography (PGC) have been utilized in describing the thermal properties of inorganic and organometallic polymers. Because of the variety of bonding energies present in many metal-containing polymers, "stability plateaus" occur where degradation followed by TG occurs through several somewhat distinct steps being seen as temperature ranges where little or no loss of weight occurs compared with similar temperature ranges where weight loss is more rapid. We have utilized this to advantage in evaluating thermal techniques since degradation pathways may be more clearly defined. Counter, the presence of low-lying filled and unfilled "d"-orbitals, present for most metals, permits a myriad of optional pathways for thermal activation followed by evolution, crosslinking and/or rearrangement to occur.

Elements such as phosphorus, silicon, tin, germanium and sulfur form catenated polymers similar to carbon, but such catenation does not usually lead to (homo) chains greater than

0097-6156/83/0229-0025$06.00/0
© 1983 American Chemical Society

10. Further, such products might be expected to (and in fact do) offer poorer thermal stabilities than carbon-based polymers since their bond energies are generally significantly less (for instance the C-C bond energy for typical hydrocarbons is about 82 kcal/mol; S-S bond energy is about 58, Si-Si is 53 kcal/mol, P-P is 48, and Sn-Sn is 39 kcal/mol). The alternative of utilizing heteroatom back-bones is attractive since the resultant bonds can exhibit greater bond energies than the C-C bond (for instance, P-N has a bond energy of 138 kcal/mol; Be-O is 124, B-O is 113, Si-O is 108, Si-N is 104, B-C is 89, and P-S is 82, exhibiting %-ionic bond characters of ~6 for B-C to 55 for Sn-O; many of these bonds have π-bond contributions). The often relatively high ionic bond character can strongly influence polymer behavior through chain and atom interactions such as π-bond overlaps, chain stiffness, ease of undergoing rearrangement reactions and ability to undergo associative reactions involving "low-lying" d-orbitals.

In this chapter macromolecules whose backbone or connective bridge contain metal-containing moieties which exhibit sufficient covalent character so as to possess directional bonding will be considered. Thus polymers such as polysodium acrylate will be omitted since the metal is bound through largely ionic forces.

Generalizations

Generalizations are appearing as the number of thermally related studies are increasing. Following is a listing of these general trends along with a brief discussion. The term "classical polymers" refer to polymers as polystyrene and polyethylene.

1. Thermal properties are approximately independent of molecular weight after a $\overline{DP}$ of about three. This is unusually low in comparison with many more classical polymers where such thermal properties as T_g, T_m and thermal degradation may vary through $\overline{DP}$'s of 100.

2. Most thermograms show stability plateaus - the result of the presence of significantly different bond energies within the polymer backbone or side chain. More classical polymers generally do not exhibit such plateaus with "complete" volatilization occurring within a few degrees after intial weight loss occurs. Such stability plateaus may be roughly divided into two categories - kinetically independent plateaus which are independent of heating rate and time and kinetically dependent plateaus which are dependent on both heating rate and time. As a crude rule -those plateaus which are parallel to the temperature or time axis are kinetically independent whereas those which slope downward are kinetically dependent. In more classical polymers the kinetically dependent behavior has been utilized to determine such factors as specific rate constants and rates

of volatilization, order of degradation and energy of activation, but because of seemingly greater complexity of organometallic polymer this has thus far not been undertaken.

3. Organometallic polymers generally undergo a darkening of color (often to a brown to black to gray) in the 300 to 600°C range, becoming lighter (often white) about 900 to 1200°C. The coloration may result from formation of aromatic intermediates. The subsequent light color corresponds to the color of the metal oxide (in air), normally white.

4. Infrared spectral bands broaden for the residues as degradation proceeds. Some new bands form--some which are sharp indicating formation of preferred intermediate residues.

5. Initial degradation is often similar in air and under inert conditions consistent with initial degradation being nonoxidative. Subsequent degradation is dissimilar and consistent with oxidatively degradation occurring in air.

DSC appears to be a more precise indicator than TG when comparing degradation routes. Thus many TG thermograms appear similar in air and under an inert atmosphere yet the DSC thermograms are markedly different.

6. A number of organometallic polymers undergo initial degradation (with or without associated weight loss) at relatively low temperatures (50 to 200°C range) but with retention of major portions of the organic portion to greater than 400°C. This is generally described as the product's having poor to moderate low temperature stabilities and moderate to good high temperature stabilities.

7. The initial thermally induced transitions often affect material processing properties as T_g, T_m and solubility and often involves rearrangement about the metal atom. This is disappointing and points to the often fruitless attempts to obtain oxidatively stable materials through shielding the metal from oxidation. Alternate attempts might emphasize procedures for "fixing" the geometry about the metal atom. Such "fixing", though, typically results in chain stiffening and subsequent poorer processing properties.

The afore is a consequence of a) often high ionic-character of the metal-associated bonds and b) presence of low-lying, available d-orbitals or other hybridized orbitals both permitting ready arrangement about the metal atom.

8. Metal-containing moieties typically act as chain stiffening agents increasing T_g and T_m but decreasing solubility. Flexibility is imparted by the hydrocarbon portions.

9. There does not appear to be a direct correlation between thermal, oxidative and hydrolytic stabilities, thus products must be designed to emphasize the separate desired property.

Hydrolytic stability is often inherent in many organometallic polymers due to the hydrophobic nature of the organic portions as long as the chains are not "wetted". For instance tin polyesters are stable to boiling water for a week, yet rapidly degrade when dipolar aprotic liquids as DMSO are added.

10. Many organometallic polymers undergo initial thermal degradation through solid transitions without melting.

11. While some organometallic polymers exhibit truly outstanding weight retention, it must be remembered that sometimes most of the remaining residue is inorganic (specifically in air -the metal oxide) thus care should be exercised when evaluating weight retention data to note how much is truly organic. Thus one should be continually aware of the meaning of weight retentions, particularly with metal-containing products which can oxidize at lower temperatures leaving not a desired moldable and/or flexible product but rather a powdery metal oxide. Even in an inert atmosphere the metal-containing moiety generally is one of the last to volatilize. For instance, the Schiff base coordination polymers of 5,5'-p-phenylene-bis-(methylidynenitrilo)di-8-quinolinol show reasonable weight retention to 400 to 500°C with only about 15-35% weight losses, but on considering the metal content this represents 50 to 75% loss of the organic portion of the polymer (2). In fact, we have found that for the majority of condensation polymers made by our group, the metal remains and in many cases thermal degradation can be utilized in determining the amount of metal present in the original polymer (1).

12. Most organometallic polymers exhibit equal to better weight retention under inert conditions compared to air.

Degradation Pathways

A number of degradation mechanisms have been identified. Some of these mechanisms are briefly described in this section.

Rearrangement reactions are one large class of degradation mechanisms including reorganization of groups about the metal atom and ring formation. Inorganic monomers tend to form small, unstrained rings as in the case of siloxanes which typically undergo extensive rearrangements at temperatures in excess of 350°C, often forming cyclic products which may involve an equilibrium between the cyclic product and the polymer. Gee (3) has treated such equilibrium as follows for an equilibrium between a short chain or cyclic product R, a shorter chain C and a longer chain composed of C + R, CR.

$$C + R \rightleftarrows CR$$

At equilibrium $\Delta G = 0$. For spontaneous processes ΔG must be negative and at constant temperature $T\Delta S > \Delta H$. The equilibrium constant is

$$K = \frac{[CR]}{[C][R]} \quad [1]$$

and where the mixture is mainly polymer [CR] = [C] and

$$K \cong 1/[R] \qquad [2]$$

The equilibrium constant is related to the standard free energy change

$$\Delta G° = -RT \ln K = RT \ln [R] \qquad [3]$$

If $[R]_o$ is the concentration of cyclic ringed products in pure cyclic product, then

$$\Delta G = \Delta G^o - RT \ln [R]_o \quad \text{and} \qquad [4]$$

$$\Delta G = RT \ln [R]_o \qquad [5]$$

In the absence of solvent $[R]/[R]_o = 1 - \Phi$, where Φ is the weight fraction of chain polymer in an equilibrium mixture.

The relationship $\Delta G = \Delta H - T\Delta S$ reduces to

$$-R \ln (1-\Phi) = \Delta S - \Delta H/T \qquad [6]$$

Three situations are derived from this equation. First, if ΔH is negative (exothermic reaction), the polymer concentration, [CR], decreases as temperature increases. If ΔS is also negative, no polymer can exist above a ceiling temperature. If ΔS is positive or zero, polymer can exist at any temperature (below which other alternative reactions occur).

If $\Delta H = 0$, the [CR] is independent of temperature and is given by $-R \ln (1 - \Phi) = \Delta S$. If ΔS is negative, the polymer will be unstable with respect to formation of R.

If ΔH is positive (endothermic), [CR] will increase as temperature increases. If ΔS is positive, polymer will not exist below a lower temperature given by $T = \Delta H/\Delta S$. If S is negative, no polymer can exist regardless of temperature.

Thus predictions can be made with regard to the effect of temperature on such equilibrium reactions given determined ΔH and ΔS values.

Polyphosphazenes typically are of the first type while sulfur and selenium are of the last type (regarding ΔH values).

The reaction of $Be_4O(OCOCH_3)_6$ with diacid chlorides (4) gives low molecular weight polymers. On heating, interchange occurred and $Be_4O(OCCH_3)_6$ could be sublimed at 110-140°C at 10^{-2} torr pressure for the product derived from adipyl chloride. The use of terephthalyl chloride appears to increase the stability of the product such that a temperature of 340°C was required to obtain sublimed beryllium acetate. Interchange also slowly occurs at room temperature.

Elimination can occur leading to the formation of stable residues or may signal the onset of continued, further degradation. For uranyl polyesters (1) elimination of "inner-core" waters at about 150°C typically leads to the formation of (2) accompanied by a change in geometry about uranium from an octahedral bipyramide to an octahedral structure (5). Residue 2 is stable to about 450°C (6). Pyrolysis of manganese polyamines (3) begins with elimination of the pyridine moiety about 250°C. This is followed by a massive breakup of the diamine moiety and elimination of water (6) without generation of a stable, identifiable, isolatable residue.

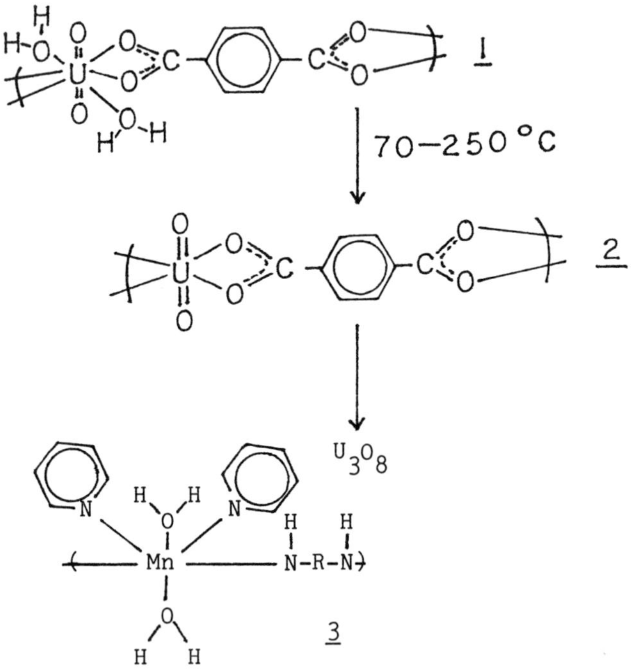

Elimination reactions leading to further degradation is by far the more common type of elimination degradation pathway.

Many organometallic monomeric compounds undergo thermally induced disproportionation. Disproportionation also occurs in metal-containing polymers. Aryloxy polyaluminum oxides(4) disproportionate on heating, forming aluminum oxides (7).

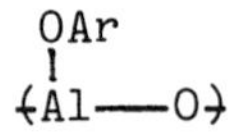

4

Unzipping pathways have been cited as a possible initial degradation pathway (1).

Oxidation of the metal atom typically occurs at some point during degradation processes occurring in air (1). Thus for tin polyamines, entrance of oxygen to the tin atom through an associative pathway with subsequent degradation involving evolution of both fragments derived from the organic backbone and organic groups bonded to the tin is believed to occur in air (1).

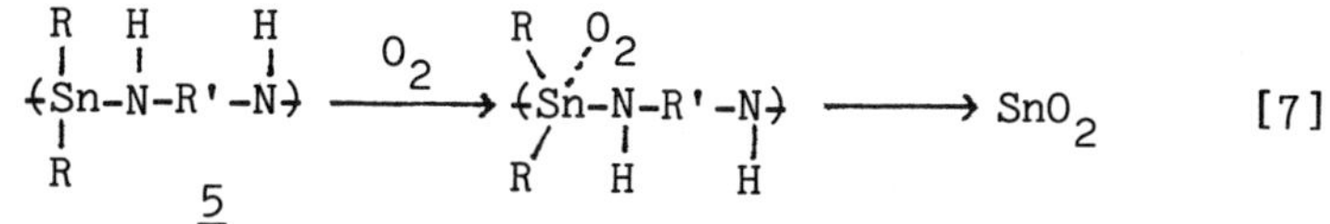

5 6 [7]

It is interesting to note that there is evidence that for some situations oxidation occurs away from the metal where the metal is connected through "ether" oxygens. For instance for alkyl silicones, the alkyl groups are oxidized typically about 250-350°C (8). Thus the situation with regard to the site of original oxidation is complex and not settled.

Oxidation often results in the formation of crosslinking through oxygen bridges. Such products can be fairly stable. Thus polyphenylsiloxane loses 8% of its weight to 400°, an additional 40% to 450°C and an addition 8% to 550°C (9). Polyvinylsiloxane loses 5% at 250°, an additional 12% to 350° and an additional 5% to 550°C. Such silanes are believed to undergo oxidative degradation initially through cleavage of the organic groups with crosslinks formed through siloxane bonds, with the new structure resisting further oxidation.

When the metal is bonded through nonoxygen bonds the results in air are such that initial degradation may involve any number of pathways and sites and is dependent on the particular polymer.

Formation of crosslinks, typically accompanied by elimination of associated moieties, is a common occurrence and serves as the basis for the formation of graphite, carbon and boron-containing fibers. This is particularly true when nonaromatic double and triple bonds are present such as 7 or where aromaticity can be gained through bond relocation.

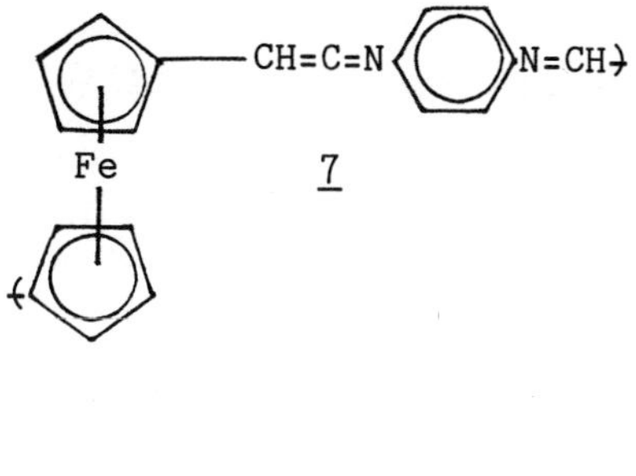

7

Borazole (8) undergoes a complex polycondensation above 500° yielding

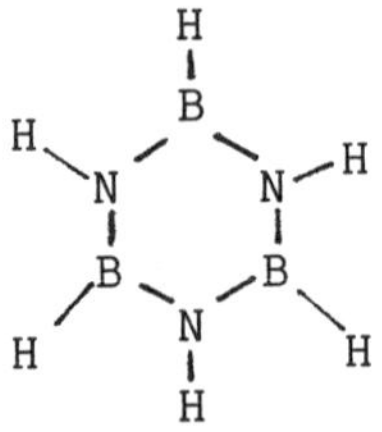

8

a BNH highly crosslinked structure (11) and polybutyl titanate also decomposes to a complex, highly crosslinked titania network. These networks are often then heated to several thousand degrees further developing the cross-linked network and leading to materials which exhibit both outstanding thermal stabilities and strengths.

Degradation through attack of the organometallic polymer by impurities is probably one of the most prevalent modes of polymer degradation, yet a mode where little has been reported.

Extensive studies carried out on polysiloxanes show that the thermal properties are also dependent on such items as chain length, endgroup, and source (i.e., nature of synthesis). The latter is related to inclusion of impurities, amount and distribution of crystallinity, etc. The presence of minute amounts of impurities can give misleading results with respect to most physical and chemical properties including thermal stability.

The chemical and physical, including thermal, resistance to degradation of polysiloxanes is a consequence of both the high Si-O bond energy (106 kcal/mol) and the relatively large amount of ionic character within the Si-O moiety. The ionic character of the Si-O bond also facilitates acid-and base-catalyzed rearrangement and/or degradation reactions. Thus highly purified polymethyl siloxanes are stable to 350-400°C where siloxane bond interchange to form cyclic products occurs (12). Yet in the presence of sulfuric acd, silicone rubber degrades at room temperature (13).

Thus the presence of trace amounts of nucleophilic or electrophilic agents can lead to a dramatic decrease in the thermal stability of organometallic polymers. This becomes increasingly important when it is remembered that many polymerizations employ such reagents as catalystic agents and even less recognized is the presence of such impurities in the monomers themselves.

One of the most serious problems facing polymer scientists is the elimination of trace amounts of water. This is particularly true with metal containing polymers which are particularly susceptible to hydrolysis. For many organometallic polymers, stability to water results from the hydrophobic nature of the organic portions. As previously noted, rapid hydrolysis occurs when the polymers are wetted by addition of liquids which are compatible with the electronic nature of the polymer chain. The activation energy for penetration of the "organic shield" by water is not large and typically achieved below 200°C for most organometallic polymers and such degradation may be responsible for the small amount of degradation occurring below 200°C observed for many organometallic polymers. Counter to this, we have observed that the thermal stabilities of many condensation organometallic polymers is independent of the mode of synthesis (i.e. employing aqueous present and free reaction synthetic systems). Even so, we have not rigorously kept the polymer from the air so that water may be attracted to the surface of the polymers. This area should be studied further.

Thus there exists a variety of initial degradation pathways. Further, it is possible that several of these can occur almost simultaneously.

More in-depth studies should be helpful in identifying the least stable site(s) allowing modification of these polymers at that site. Thus metal polyesters derived from 1,1'-dicarboxyferrocene generally degrade in air at about 250°C. The degradation begins with a splitting out of CO_2 and the ferrocene moiety. The degradation is rapid, possibly explosive, with as great as 70% weight loss occurring over a 10°C range (14). The analogous polyethers typically are stable to about 300°C where a more gentle degradation occurs. Thus replacement of the ester group, which readily forms CO_2, for an ether grouping allowed for the synthesis of more thermally stable polymers.

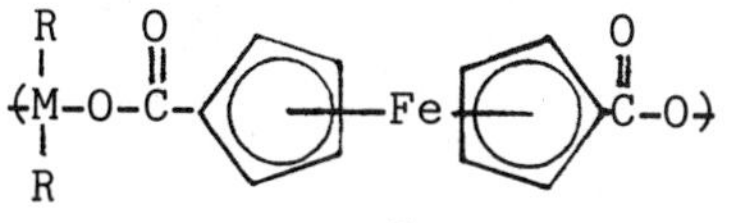

9

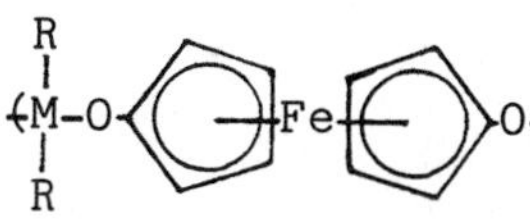

10

Attempts to Enhance Thermal Stabilities

As previously noted, siloxanes undergo extensive rearrangements at temperatures in excess of 350°C, often forming cyclic products. Because of the significant mechanical, chemical, and electrical properties offered by siloxanes and modified siloxanes, great effort is still directed towards increasing the use temperature of such products. This work is closely associated with sophisticated thermal analysis systems and is often aimed at preventing the depolymerization (reversion) reaction.

One approach often employed for preventing reversion is modification of the siloxane. Several recent attempts are described below.

Reversion and ring formation have been partially overcome through placement of chain-stiffening units in the silicone-backbone. Thus linear D_2-m-carborane-siloxanes (11-14) with one to three trifluoropropyl moieties per repeating unit) exhibit better thermal and oxidative stability than silicones and fluorosilicones (15). Initial degradation occurs in air about 300 to 350°C, almost 100° above that typically experienced for siloxanes and fluorosilicones. The carborane-siloxanes exhibit T_g's from -50 to 0°C (15). The carborane moiety also acts to inhibit formation of six-membered rings because of its size.

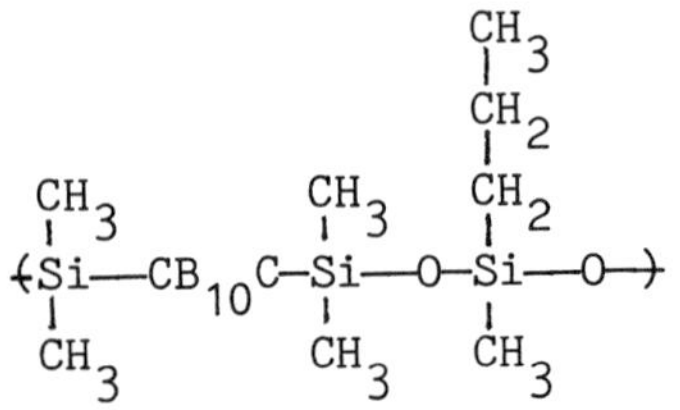

Ucarsil F_1

11

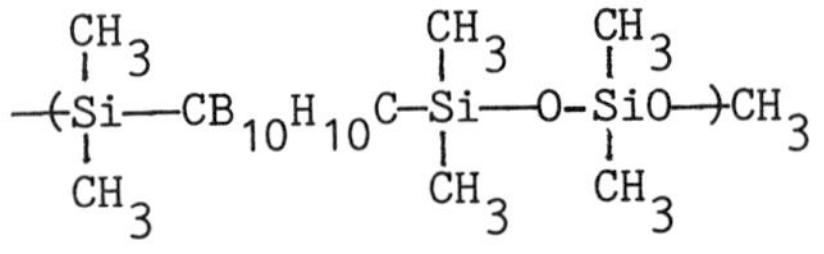

Ucarsil Me_2

12

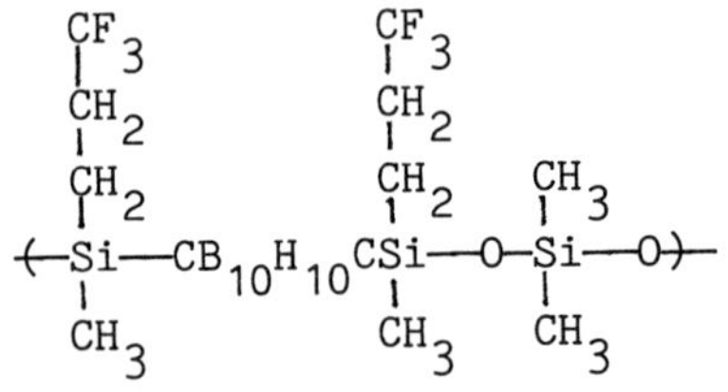

Ucarsil F_2

13

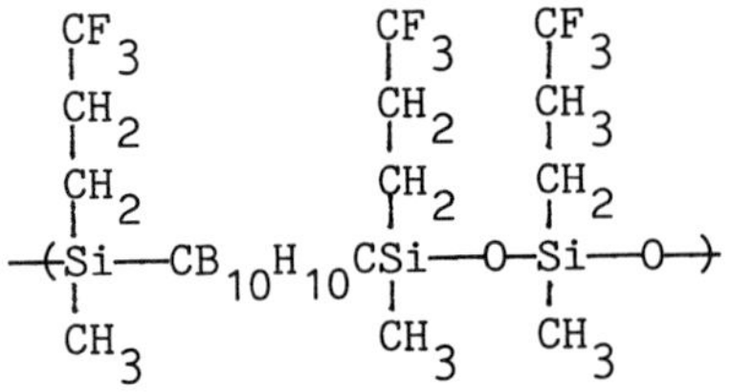

Ucarsil F_3

14

Thermal analysis of a number of linear polycarboranesiloxane copolymers and block polymers (such as 15-17) including TG, DT, TB, and TM was made by Roller and Gillham (16). A number of physical transitions were reported including crystallization temperature (10 to 230°C), glass transition temperature (-108 to 25°C), glassy-state transition temperature (-145 to -90°C), and melting temperature (40 to 260°C). This is one of the most complete single papers on the thermal analysis of inorganic polymers regarding the types of thermal properties cited.

A number of results were found that may have otherwise gone unnoticed in briefer studies (16). The cyclization mechanism of degradation which occurs for polydimethylsiloxane is not operative for polycarboranesiloxanes. At high temperatures the polycarboranesiloxanes stiffen, possibly due to crosslinking. In argon the polycarboranesiloxanes generally showed an increase in thermal stability of about 50°C over that of polydimethylsiloxane with many being stable at 500 to 550°C.

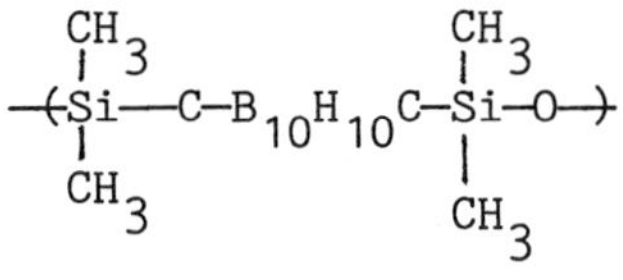

15

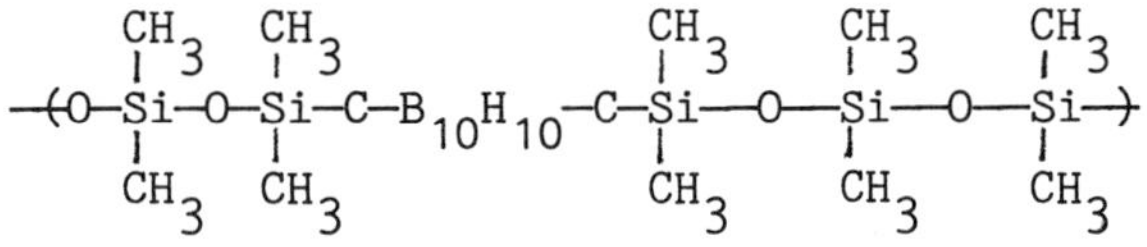

16

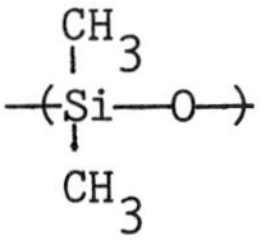

17

Another utilized approach for preventing reversion involves crosslinking of the siloxanes. For instance, a number of methyl silicone rubbers were crosslinked, (about 1 crosslink per 3,000 to 20,000 units (10-21) for crosslinked products mainly composed of dimethylsiloxane; Refs. 17 and 18). DSC and TM was utilized to evaluate the thermal properties of the products. TM was utilized to determine crosslink density along with typical modulus properties. Both enhancement and decrease in thermal properties were observed depending on the mode and conditions of crosslinking.

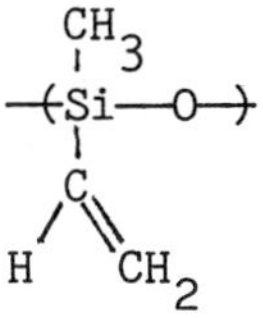

18

and

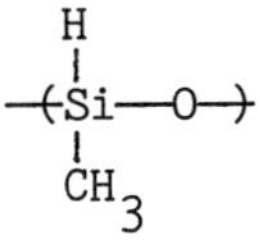

19

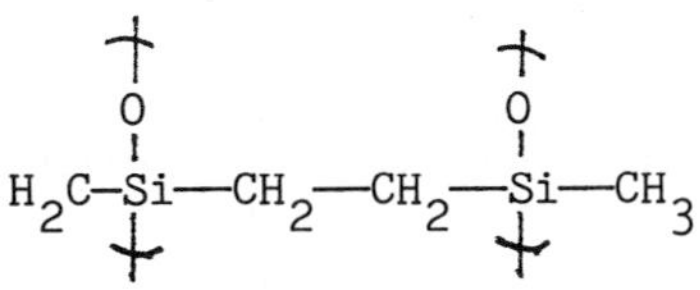

20

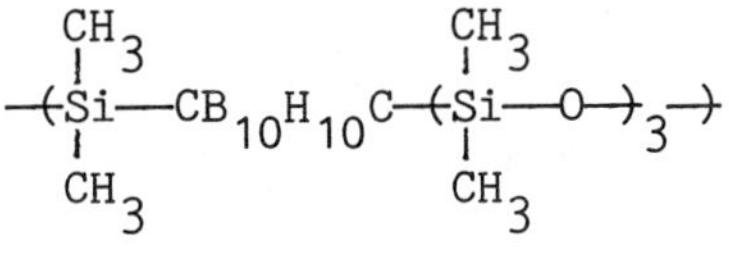

21

Another approach to combat reversion is the placement of the siloxane or modified siloxane on an "inert" support which can assist in maintaining the original nonreverted structure. Thus Finch (19) reported the formation of a new chromatographic phase stable to 500°C based on a polycarboranesiloxane stationary phase. The system was capable of efficient separations from 50 to 500°C. In fact, TG's of the polycarboranesiloxane phase showed complete weight retention to about 700°C. The meta-carborane was utilized in this study.

Siloxane-ladder polymers have been known for some time. A polymer derived from phenyltrichlorosilane (2), while infusible, is soluble in solvents such as tetrahydrofuran and benzene from which films can be cast (20,21). TG thermograms in air show no weight loss to 525° C but an upper limit for useful application is probably near 300° C (21).

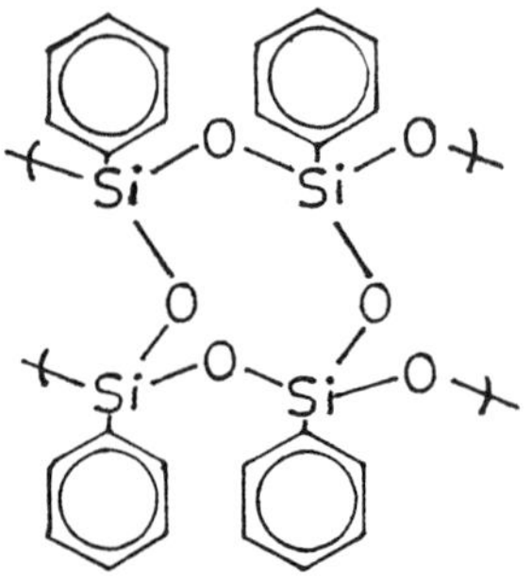

22

Several researchers have successfully employed the addition of antioxidants to improve thermal stability. Thus Nannelli, Gillman and Black (22) found that the degradation of 23 proceeds in air through oxidation and cleavage of the organic side groups with a degradation inception temperature around 200°C. The gas chromatograms of the volatile decomposition products of $Cr[OP(CH_3)(C_6H_5)O]_2OP(C_6H_{17})_2O$ showed at least 14 products of which 2-octane, acetic acid, and n-butyric acid are among the most abundant. Decomposition does occur at 200°C for long exposure times with a 13% weight loss for an 18-hour exposure time. Addition of antioxidants such as distearyl thiodipropionate increases the longevity of the polymer films flexibility at 200°C.

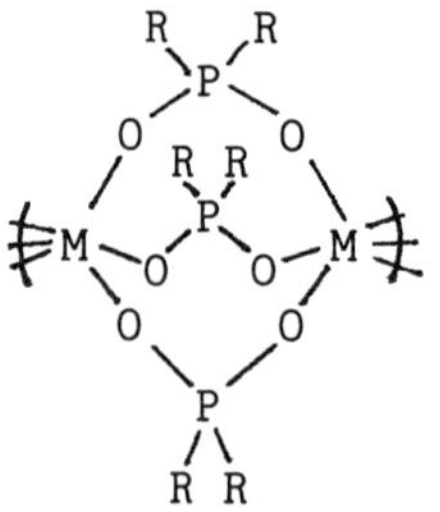

23

Differentiation of Depolymerization Mechanisms

MacCallum (25) related the change in number average molecular weight, $\overline{DP}$, with depolymerization mechanism dividing the pathways into the groups characterized by the nature of the initiation reaction-random scission or end group. These groups can be further divided into initial degradation followed by a) partial unzipping and b) complete unzipping.

Polymer degradation can be described in terms of the weight of polymer, W, at time t, number of molecules, N, at time t, and $\overline{DP}$ as follows where m is the molecular weight of each mer unit.

$$W = N\, m\, \overline{DP} \qquad [8]$$

Differentiation with respect to time gives

$$m^{-1} \frac{dW}{dt} = N \frac{d\overline{DP}}{dt} + \overline{DP} \frac{dN}{dt} \qquad [9]$$

For random initiation followed by incomplete unzipping and assuming that depolymerization is first-order with respect to sample weight gives

$$dN/dt = kW/m \text{ and} \qquad [10]$$

$$-dW/dt = kWZ \qquad [11]$$

where Z is the average unzipping length, i.e. number of mers separated from the chain. Combination of [8]-[11] followed by integration gives

$$\overline{DP}/\overline{DP}_o = \frac{Z}{\overline{DP}_o}\left(\frac{1-C}{C\,\overline{DP}_o + Z}\right) = f\ (1-C)/(C+f) \qquad [12]$$

where $f = Z/\overline{DP}_o$ and C, the fractional conversion, is $(W_o-W)/W_o$.

For random initial depolymerization followed by complete unzipping, Z can be related to $\overline{DP}$ through a parameter, B, that is related to the polydispersity of the polymer sample.

$$Z = \overline{DP}\ B \qquad [13]$$

Combining [10], [11] and [13] into [9] with integration gives

$$\overline{DP}/\overline{DP}_o = (1-C)^{(B-1)/B} \qquad [14]$$

For chain-end initiation followed by partial unzipping

$$-DN/dt = 0 \qquad [15]$$

and from [9], [11] and [15] with integration

$$\overline{DP}/\overline{DP}_o = 1-C \qquad [16]$$

For initial end-group depolymerization followed by complete unzipping

$$-dN/dt = kN \qquad [17]$$

and combining [8], [9], [11] and [17] followed by integration yields

$$\overline{DP}/\overline{DP}_o = 1 \qquad [18]$$

In a situation where intramolecular "loops" of constant size are formed and rapidly removed (as volatiles under reduced or ambient pressure) equations [11] and [15] apply and can be incoporated into [9] giving [16] after integration.

Thus for end-group depolymerization, $\overline{DP}$ is a linear function of time for partial unzipping and independent of time for complete unzipping. For random bond breakage partial unzipping is a complicated logarithmic function of $\overline{DP}$ related to reaction extent whereas complete unzipping is related to a linear function of $1/\overline{DP}$ with time.

The afore situation can be pictured as shown in Figure 1. Line A, corresponding to initial end-group depolymerization followed by complete unzipping shows no change in molecular weight since the initiation of depolymerization is the rate-determining step;i.e. chain lengths of remaining chains remain unchanged. Counter, if end-group depolymerization is followed by incomplete unzipping, C, average chain length will decrease slowly, the change in average chain length related to the size of segment removed, including loops. A curve such as B results where random sission is followed by complete unzipping since there is a relatively slow decrease in molecular weight with conversion. Random depolymerization followed by incomplete unzipping results in a rapid decrease in molecular weight with conversion and is pictorialized as D. Both B and D are members of a number of similar curves, the exact shape being dependent on the exact depolymerization parameters. These curves can be computer-generated for varying situations.

Degradation may involve more than one pathway. The afore approach can be extended to consider these cases and others. Thus for degradation occurring through formation of volatile loops as well as through an internal, intermolecular process,

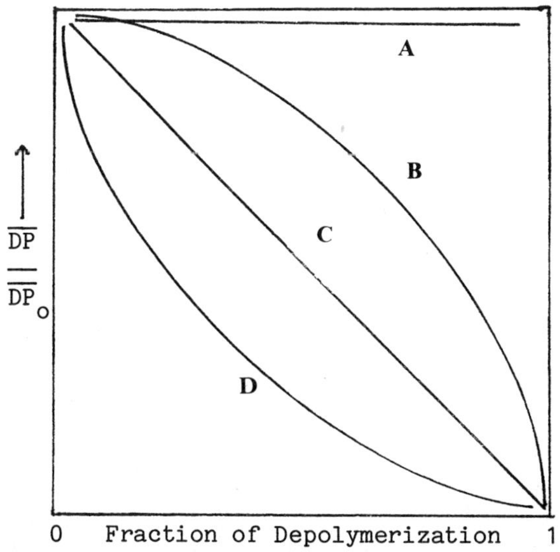

Figure 1. Idealized plots relating molecular weight ratio to depolymerization mechanisms. Key: A, end-group depolymerization followed by complete unzipping; B, random scission followed by complete unzipping; C, end-group depolymerization followed by incomplete unzipping; and D, random depolymerization followed by incomplete unzipping.

equations [8] and [9] still hold. The rate of weight loss is given by

$$-dW/dt = kmZW = k'W \qquad [19]$$

The rate of change of molecular population in the condensed phase depends on the relative magnitudes of the rate of loop formation forming volatile and nonvolatile oligomers, etc. These changes can be assumed to be a weak function of time as follows where A(t) describes the time dependence of the particular systems.

$$dN/dt = A(t) \qquad [20]$$

This function can be parametized in a number of ways including involving initial population, polymer lifetime and other sealing parameters. Models combining growth and decay through various routes can be made and computer plots made with ensuing curve fitting.

Thus there exists several ways to determine the mode of depolymerization with most requiring the products to be soluble permitting molecular weight to be determined. Further these approaches are best suited to situations where depolymerization occurs through a single mechanistic pathway. Thus for polymers as siloxanes the afore techniques should be readily applicable but for many metal-containing polymers degradation proceeds through the formation of insoluble residues.

Coupled Instrumentation Employing MS

PY-GC-MS and GC-MS have been extensively utilized in both traditional product identification and in select studies to assist in the description of the thermal degradation of the materials. These techniques typically do not allow a ready, straightforward interpretation of degradation sequences due to considerable combination of evolved fragments on the GC columns.

In an effort to overcome problems of combination, we first constructed a TG-MS assembly where the evolved species from thermodegradation were swept, by a helium purge gas, in a straight line where the distance between the TG sample compartment (boat) and the MS ionization source was about three feet (23). While numerous ion fragments were found, it was surprising that certain species still underwent combination. This was found for phenyl groups derived from a number of different samples. The amount of biphenyl formed was in great excess with that calculated using simple gas collision theory. Presumably combination is occurring within the solid sample prior to evolution of that fragment. Formation of biphenyl occurred even when the samples were ground to a fine powder.

The TG-MS assembly consisted of a double-focusing DuPont 21-491 Mass spectrometer coupled through a single-stage glass jet separator to a DuPont 951 Thermal Gravimetric Analyzer which was attached to a DuPont 990 Thermal Analyzer Console. The MS was equipped with a Hewlitt-Packard, HP-2216C computer having 24K core memory and a disc-oriented data system specially developed for the DuPont 21-491 Mass Spectrometer. The MS system can be controlled by the computer system, which includes a dual 2.5M byte disc drive, a Hewlett-Packard Cathode Ray Tube terminal, a Tektronix storage scope (for display) driven by a dual 12 bit digital-to-analog (D/A) converter, and Versatec printer/plotter. Data was acquired using a 14 bit analog-to-digital (A/D) converter (13 plus sign). This system can operate and process data at rates up to 8 KHz.

A thermocouple, interfaced with the Hewlett-Packard 2116C computer, was employed to sense the temperature during the course of the thermogravimetric analysis. This probe was attached to the TG in order to monitor the actual temperature of the sample during each run. During sample analysis, the TG-jet separator connection was wrapped with heating tape and maintained at a temperature of 125°C while the jet separator oven was maintained at 135°C. The MS-jet separator transfer line was also wrapped with heat tape and maintained at 165°C while the source temperature was held at 230°C.

The assembly and calibration procedures are described in detail in reference 23. Depending on the heating rate, up to several hundred complete spectra are obtained as a function of temperature with the data available as a function of absolute ion abundance and normalized ion abundance for each ion such that plots of ion intensity as a function of temperature can be made for all mass numbers.

A number of metal-containing polymers were evaluated employing the TG-MS assembly. Organometallic polymers were chosen for several reasons. First, since degradation of these compounds typically occur by different routes in air and in inert environments, the two environments can be clearly, easily differentiated. Second, some of the organometallic polymers exhibit good high temperature stability, losing less than 20% of their initial weight to the 800-1200°C range. Identification of the degradation products from the polymer would be useful in better understanding their good stability, and in designing and synthesizing still more stable polymers. A third reason for using organometallic polymers in assessing the TG-MS is that the degradation of these materials occurs in a stepwise manner over a wide temperature range (often over a range in excess of 600 C°) with interspersed "stability plateaus". Continuous monitoring of the evolved chemical products, as is possible with the TG-MS described, is useful in understanding these transitions in the degradation sequence. Finally, the organometallic polymers are good candidates for study by TG-MS because they are difficult to characterize by other techniques. Conventional C, H, N elemental analysis often yields poor results, and since many of these polymers are only sparingly soluble in most solvents, characterization by NMR or ESR is not practical. Methods such as those described herein are therefore needed for characterizing the structures of such organometallic polymer materials.

A number of titanium polyethers derived from hydroquinone derivatives were evaluated employing the TG-MS assembly and infrared spectroscopy (24). Two general trends were found. First, the initially evolved degradation product was cyclopentadiene typically appearing about 100°C. Second, this was followed by evolution of the hydroquinone moiety along with the associated oxygens. There was great variation with respect to the extent and rate of degradation of the hydroquinone-containing moiety from the several polymers studied, and the quantities of organic components in the final residues from degradation also varied markedly.

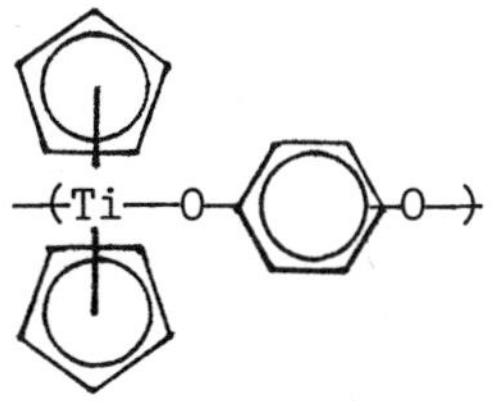

24

The afore suggests that the limiting thermally stable moiety is the cyclopentadiene moieties and that construction of thermally stable titanium-containing polymers should avoid the Cp_2Ti moiety.

At a heating rate of 20 C°/min, the cycle time for sample is in excess of one hour making such determinations commercially costly.

More recently we have developed a similar assembly except employing a programmable PY in place of the TG (14). Thus TG are obtained on samples to determine appropriate temperatures for MS to be obtained. Briefly, uranyl polyesters show three somewhat distinct stability plateaus. The PY was programmed to heat the samples to the inception of the stability plateaus plus 50 to 100C°. The MS obtained at 250 to 300°C indicated evolution of water accounting for a 5 to 10% weight loss. At 450°C mass fragments derived from CO_2 were found. At 900°C mass fragments derived from the vinyl group were found.

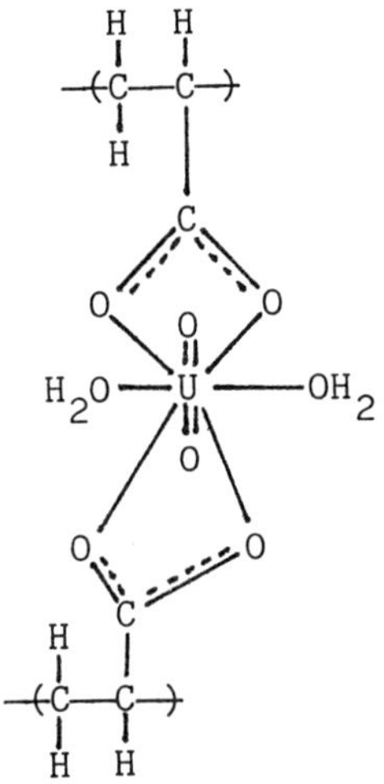

25

The cycle can be accomplished within 15 mins. making it commercially more acceptable.

Summary

Few definitive thermal studies of metal-containing polymers (with the exception of silicon-containing polymers) exist in the literature. Thermal analysis is typically done as a matter of preliminary testing on new polymers. Further, most of the studies were done prior to the advent of the automatic, programmed thermal analysis instruments.

The lack of definitive studies is due to a mixture of reasons including 1) wide variety of polymers; 2) newness of interest in the area; 3) wide variety of applications (both potential and actual) of inorganic and organometallic polymers not requiring thermal stability or thermal analysis (uses as anchored metal catalysis, control release agents, electrical and photochemical applications, speciality adhesives); 4) insufficient description, identification, of the products; 5) wider variety of degradation routes and other thermal behavior in comparison to organic polymers; and 6) many products were synthesized and briefly characterized before the advent of modern thermal instrumentation.

The thermal techniques that have already been utilized on more classical polymers are generally directly applicable to organometallic and inorganic polymers. The thermal analysis of inorganic and organometallic polymers is both more and less difficult in comparison to studies performed on organic polymers. On the negative side there are 1) a much wider variety of products with many polymers showing a great dependence of the thermal properties on the presence and nature of impurities; 2) general lack of definitive thermal studies and experience upon which to build; and 3) added difficulty associated with defining thermal responses. On the positive side the wider variety of bond energies present in organometallic and inorganic polymers can allow (for certain products) a clearer isolation and definition of particular thermal transitions, such as T_g's and degradation pathways. Further, the wider variety of thermal response permits increased potential usefulness of such compounds showing unusual (either in extent, range, or actual thermal response) thermal responses.

Recommendations

There is a need both for studies involving a wide variety of products and in-depth studies of several of these systems. The "survey" studies can indicate potentially interesting polymers such as products showing unusual and/or potentially useful thermal responses.

The correlation of bonding group, metal atom, and fine and gross polymer structure with such factors as degradation pathway, bonding energies and inherent stabilities should begin.

There is a need to evaluate, by newer thermal analysis instrumentation and techniques, polymers previously only briefly characterized, emphasizing those products which show potential industrial application or other meritorious property. Such products should be well characterized with regard to such factors as chain length, molecular weight distribution, endgroup, purity, nature and amount of impurities, and actual morphological structure of the polymer.

The increased application of modern thermal techniques and companion techniques must occur if the thermal characterization of organometallic and inorganic polymers is to progress. A number of companion techniques have been shown to be useful and merit further use. These include NMR, IR, MS, GC-MS, AA, X-ray microscopy, and GPC. ESR has not been appreciably utilized as a companion technique but may be considered, particularly with polymers containing ESR-sensitive metals. Gel permeation chromatography (GPC) should be considered as a companion tool for thermal degradation evaluations where depolymerization is suspected. There is a need for a number of well-chosen thermal analysis studies of both the volatile products by TG-MS, etc. (for both chemical and toxicological purposes) and the residue. Such studies should occur throughout the studied temperature range rather than just at the room and final temperatures.

Acknowledgment

The author thanks Martel Zeldin for his help in preparing the section on differentiation of depolymerization mechanisms.

Literature Cited

1. C. Carraher, J. Macromol. Sci.-Chemistry, A17(8), 1293 (1982).
2. E. Horowitz, M. Tryon, R. Christensen and T.P. Perros, J. Appl. Polym. Sci., 9, 2321 (1915).
3. G. Gee, Chem. Soc. (London) Spec. Publ. No. 15, 65 (1961).
4. C.S. Marvel and M.M. Martin, J. Amer. Chem. Soc., 619 (1958).
5. C. Carraher and J. Schroeder, Polymer P., 19 (2), 619 (1978).
6. C. Carraher, V. Foster, H. Molloy, M. Taylor, T, Tiernan and J. Schroeder, J. Macromol. Sci.-Chem., A16(1), 231 (1981).
7. R. Brotherton, Conference on High Temperature Polymer & Fluid Research, A50-TDR-62-372 (August 1962), p. 389.
8. K. Andrianov, Soc. Chem. Ind. (London) Monograph No. 13, 75 (1961).
9. K. Andrianov, Metalorganic Polymers, Interscience, NY, 1965.
10. E. Neuse, H. Rosenberg and R. Carlen, Macromolecules, 1, 424 (1968).
11. A.W. Laubengayer, P.C. Moews and R.F. Porter, J. Amer. Chem. Soc., 83, 1337 (1961).
12. C.W. Lewis, J. Polymer Sci., 33, 153 (1958).
13. A.V. Tobolsky, Inorganic Polymers, Academic Press, N.Y., 1962, pp. 10-27.

14. C. Carraher and J. Schroeder, Unpublished results.
15. E. Peters, D. Stewart, J. Bohan, R. Moffitt, C. Beard, G. Kwiatkowski and E. Hedayn, J. Polymer Sci., 15, 973 (1977).
16. M.B. Roller and J.K. Gillham, Polym. Eng. Sci., 14, 567 (1974).
17. E.M. Barrall, M. Flander and J.A. Logan, Thermochim. Actn, 5, 415 (1973).
18. P.J. James, E.M. Barrall, B. Dawson and J.A. Logan, J. Macromol. Sci.-Chem., A8, 135 (1974).
19. R.W. Finch, Analabs Res. Notes, 10(3), 1 (1970).
20. J.F. Brown, L. Vogt, A. Karchman, J. Eustance, K. Kiser and K. Krantz, J. Amer. Chem. Soc., 82, 6194 (1960).
21. J.F. Brown, J. Polymer Sci., C1, 83 (1963).
22. P. Nannelli, H. Gillman and B. Block, J. Polymer Sci., A-1, 9, 3027 (1971).
23. C. Carraher, H. Molloy, T. Tiernan, M. Taylor and J. Schroeder, J. Macromol. Sci.-Chem., A16(1), 195 (1981).
24. C. Carraher, H. Molloy, T. Tiernan, S. Tsuji, M. Taylor and S. Coldiron, Unpublished results.
25. J.R. MacCallum, European Polymer J., 2, 413 (1966).

RECEIVED January 20, 1983

SPECIFIC EXAMPLES OF POLYMERS

1982

ARTHUR K. DOOLITTLE AWARD

The following paper, "New Styrene-Maleic Anhydride Terpolymer Blends" by W. J. Hall, R. L. Kruse, R. A. Mendleson, and Q. A. Trementozzi, received the 1982 Arthur K. Doolittle Award.

The Arthur K. Doolittle Award is presented annually to the best paper presented at the Division of Polymeric Materials: Science and Technology (PMSE) meetings. The Award is $1000 and a plaque and is financed by the gift of the Union Carbide Corporation of the royalties of A. K. Doolittle's book "Technology of Solvents and Plasticizers". Papers are evaluated on the following criteria:

- Relevance to areas encompassed by applied polymer science.

- Originality of ideas, experimental techniques, and treatment of subject matter.

- Depth and breadth of investigation.

- Contribution to the advancement of scientific knowledge.

- Clarity of paper as it appears in the preprint book.

- Clarity of presentation at the meeting.

- Quality of audio-visual material.

- Audience interest at the meeting.

Recipients of the award are selected by an anonymous committee appointed by the PMSE Secretary. The first award was made in 1955. A complete list of award winners may be found in the ACS Division of Organic Coatings and Plastic Chemistry Preprints, Volume 45.

5

New Styrene–Maleic Anhydride Terpolymer Blends

W. J. HALL, R. L. KRUSE, R. A. MENDELSON, and Q. A. TREMENTOZZI

Monsanto Company, Indian Orchard, MA 01151

A group of new, fully miscible, polymer blends consisting of various styrene-maleic anhydride terpolymers blended with styrene-acrylonitrile copolymer and rubber-modified versions of these materials have been prepared and investigated. In particular the effects of chemical composition of the components on heat resistance and the miscibility behavior of the blends have been elucidated. Toughness and response to elevated temperature air aging are also examined. Appropriate combinations of the components may be melt blended to provide an enhanced balance of heat resistance, chemical resistance, and toughness.

Engineering plastics are generally developed to provide appropriate property response to increasingly severe environmental conditions; e.g., heat, stress at temperature extremes, and chemicals. A number of such polymers have been introduced in recent years, each with its own balance of property advantages and disadvantages. Simultaneously, research on polymer compatibility has significantly increased the number of known completely miscible polymer pairs (1), which generally show approximate additivity of many properties as a function of component ratio. This characteristic can, in principle, be used to create a family of materials with optimized property combinations, among which are those properties which qualify such polymer blends as engineering plastics. This paper discusses a group of new fully miscible polymer pairs consisting of various styrene-maleic anhydride random terpolymers (S/MA/X) and styrene-acrylonitrile random copolymer (SAN) and rubber modified versions of these pairs. Appropriate combinations can provide an enhanced balance of heat resistance, chemical resistance, and toughness. The termonomers in S/MA/X to be discussed are acrylonitrile (S/MA/AN), ethyl acrylate (S/MA/ETA), isobutylene (S/MA/IB), methyl acrylate (S/MA/MEA), and methyl methacrylate

0097-6156/83/0229-0049$06.00/0
© 1983 American Chemical Society

(S/MA/MM). Miscibility behavior of typical matrix pairs is discussed as a function of composition, and discussion of material behavior in this work emphasizes response to elevated temperatures and to stress at temperature extremes. Thus, glass transition temperature (Tg), distortion temperature under load (DTUL), and tensile deformation properties as a function of temperature are reported for various typical blend combinations. Also discussed are the results of elevated temperature air aging studies on a typical blend compared to ABS.

Experimental

The various copolymer and terpolymer samples, with or without grafted polybutadiene rubber, were produced by conventional free radical polymerizations. Blending was performed using one of several methods: Brabender small scale mixing bowl, Brabender 3/4" extruder, large scale Banbury intensive mixer or large scale extruder, depending on the scale of the experiment desired. In the work discussed here no effect of blending scale was observed. Tg's reported here were measured by differential scanning calorimetry (DSC), using a DuPont 990 Thermal Analyzer, at 20°C/min. heating rate and taking the onset value as Tg. DTUL measurements were performed according to ANSI/ASTM D648 with 1.82 MPa maximum fiber stress loading on unannealed specimens with dimensions given appropriately in the Results and Discussion section. Tensile deformation measurements were performed at various temperatures according to ANSI/ASTM D638 on injection molded Type I (12.7 x 3.2 x 57 mm gauge dimensions) specimens, using an Instron Universal Tester and Instron constant temperature chamber (the latter for all temperatures other than room temperature (R.T. = 23°C)). Tensile testing speed was 5.08 mm/min. (0.2 in./min.) grip separation rate in all cases. Izod impact testing was performed according to ASTM D256 on 12.7 x 3.2 x 63 mm specimens, injection molded in the case of the aging study and compression molded in the cases given in Table I. Specimens were notched to a depth of 2.5 mm with a root radius of 0.25 mm.

The elevated temperature aging study was performed on injection molded tensile, Izod, and DTUL specimens (dimensions given above) which were maintained in a circulating air oven at an accurately controlled 90°C for periods of time up to two months. When removed from the air oven, these specimens were conditioned at 23°C and 50% R.H. for at least 24 hours before testing. Notching of the Izod bars was performed on conditioned specimens.

Results and Discussion

Of major importance in the development of most engineering plastics is the retention of rigidity to as high a use temperature as is possible consistent with the optimization of such other properties as toughness, resistance to chemical attack, appearance of the fabricated article, etc. Thus, temperature, or heat, may be considered to be a hostile environment to which the engineering plastic is submitted. A wide variety of S/MA/X terpolymers were investigated to create the optimum balance of properties taking advantage of the well-known enhancement of Tg with increase of MA in S/MA. In this paper we discuss the effect of terpolymer composition on rigidity in terms of Tg and, in particular, DTUL. Figure 1 shows a plot of DTUL (in this case, 12.7 x 12.7 x 127 mm injection molded specimens) vs. MA content for S/MA copolymers (from the data of the late Dr. Y. C. Lee (2)) and, as an example of the terpolymer cases, of Tg vs. MA content in S/MA/MM terpolymer at ca. 6% (wt.) methyl methacrylate. The rate of increase of Tg with MA content is approximately 2°C/percent (wt.) MA. Figure 2 shows DTUL (12.7 x 3.2 x 127 mm, compression molded) vs. weight percent of the third monomer for the cases of S/MA/AN, S/MA/IB, and S/MA/MM, both glassy polymer and rubber modified versions, where the data are taken from Lee (2) and Lee and Trementozzi (3, 4). The methyl acrylate and ethyl acrylate cases are not shown in order to preserve clarity in the figure, but both give somewhat lower DTUL's at comparable (wt.) compositions. In all rubber-modified (R.M.) cases the percent of a component given is based on total material, and in the R.M.S/MA/AN case DTUL data for samples varying from ca. 17 to 23% MA are normalized to a constant 20% MA content. Surprisingly, in the methyl methacrylate case a broad maximum in DTUL is observed. This is consistent with Tg data, not shown here, for these materials.

Terpolymer composition also affects toughness. The particular case of S/MA/AN is illustrated in Table I, where three samples of varying AN content, but roughly comparable MA and rubber contents are described in terms of Izod impact. In this case it may be seen that increasing AN level in the terpolymer leads to a general increase in the Izod impact strength. The apparent equivalence of the 0% and 5% AN cases may suggest that 5% AN in the terpolymer is too low a level to cause a sensible effect on impact strength. Also shown are two blends of the 5.1% AN S/MA/AN terpolymer with ABS (one blend containing α-methyl styrene/AN copolymer for Tg improvement). Significant toughening is observed in these latter cases.

Blending of polymer with ABS offers certain significant property enhancement opportunities such as resistance to chemical attack which increases with increasing AN content, toughening, and modification of processability characteristics. However, such opportunities can only be realized if sufficient miscibility of the glassy (matrix) components is achieved to retain or enhance toughness; and predictable variation of properties with blend ratio is to be expected only for largely or fully miscible matrix components. For this reason, a study of the miscibility

Table I. Izod Impact Strength of S/MA/AN Terpolymers and Blends

Material	*% AN	% MA	% Rubber	Izod Impact (J/m)
R.M. S/MA	0	20.7	13.7	112
R.M. S/MA/AN	5.1	22.3	14.3	112
R.M. S/MA/AN	9.5	19.9	13.4	135
**R.M. S/MA/AN//ABS//α-MS/AN (50/38/12 by wt.)	2.5	11.1	20.0	157
**R.M. S/MA/AN//ABS (60/40 by wt.)	3.1	13.4	21.9	157

*Percentage based on total blend composition except % AN given includes only terpolymer contribution

**Blends with ABS made with 22.3% MA, 5.1% AN terpolymer

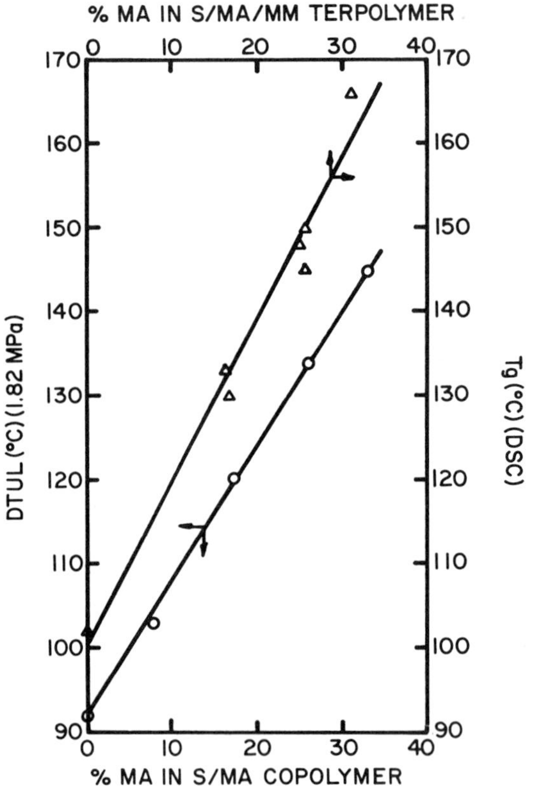

Figure 1. Effect of maleic anhydride content on distortion temperature under load (DTUL) and on glass transition temperature (Tg). Key: ○, S/MA copolymer; and △, S/MA/MM terpolymer (ca. 6% MM).

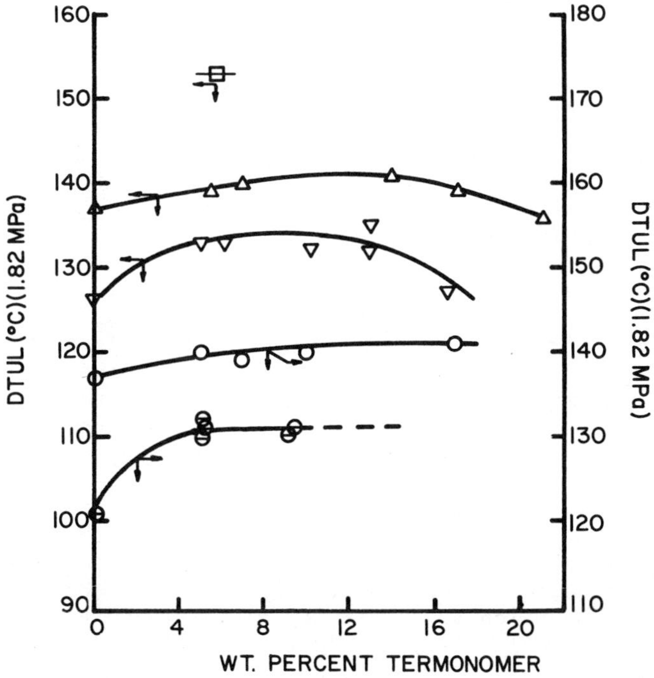

Figure 2. Effect of termonomer content on DTUL in glassy and rubber-modified terpolymers. Key: ○, S/MA/AN (25% MA); ⊖, R.M. S/MA/AN (corrected to 20% MA; ca. 14% rubber); □, R.M. S/MA/IB (29% MA; 18% rubber); △, S/MA/MM; ▽, R.M. S/MA/MM (22.5% MA; ca. 14% rubber).

of the terpolymers with SAN over a range of compositions of the components was undertaken. The particular case of S/MA/MM and SAN is discussed here as an example. Tg behavior and optical clarity were used to characterize miscibility. Thus the observation of a single Tg in a blend, combined with optical clarity, is taken to indicate total miscibility, while two Tg's identical with those of the pure components indicate immiscibility, and two Tg's at temperatures between those of the pure components indicate partial miscibility (5). Figure 3 illustrates two immiscible blend composition cases, (a) and (e), and three miscible blend composition cases, (b), (c) and (d), where the compositions of the components are given in the figure; and in all but one case five blend ratios were investigated. (The data in Fig. 3(a) are for SAN//S/MA copolymer.) It was noted for all systems studied that at any particular pair of component compositions the qualitative phase state description, i.e., miscible, immiscible, partially miscible, does not change with changing component pair ratio. Moreover, temperature dependent optical transmission experiments on a programmed hot stage gave no evidence of a phase change, i.e., lower critical solution temperature behavior, to temperatures well into the processing range for any of the (glassy state) miscible blends. Moreover, the presence of methyl methacrylate, at least to 13%, in the terpolymer does not appear to affect miscibility (see Figure 3). These observations allow a simple two-dimensional mapping of the miscibility composition region in terms of percent AN in the SAN vs. percent MA in the S/MA/MM, as shown in Figure 4, where both S/MA copolymer and terpolymers of varying MM levels are represented. As may be seen in Figure 4, a relatively wide region of miscible component compositions is available, permitting alloying of the terpolymer with ABS. This result is qualitatively consistent with the roughly similar dependence of solubility parameter on copolymer composition in the SAN and S/MA cases, as may be calculated from molar additivity rules (see (6)). Also, the solubility of SAN with itself and with polystyrene has been reported by Molau (7) and is consistent with Figure 4. Similar miscibility behavior with SAN was observed in the cases of the other terpolymers.

Blending of the high heat resistant terpolymer with ABS to form a miscible matrix phase, as defined in Figure 4, is of course, expected to increase the heat resistance of the ABS. The data in Table II, taken from references (2-4), provide a systematic representation of the variation of DTUL with terpolymer/ABS blend composition for the AN, IB, and MM-containing terpolymers (both glassy and rubber-modified). In some of these cases, as noted, a copolymer of α-methyl styrene/acrylonitrile was added to the formulation. From these data it is apparent that DTUL does, indeed, increase with increasing terpolymer concentration in the blends. Moreover, the effect of the terpolymer composition on DTUL of the blends is

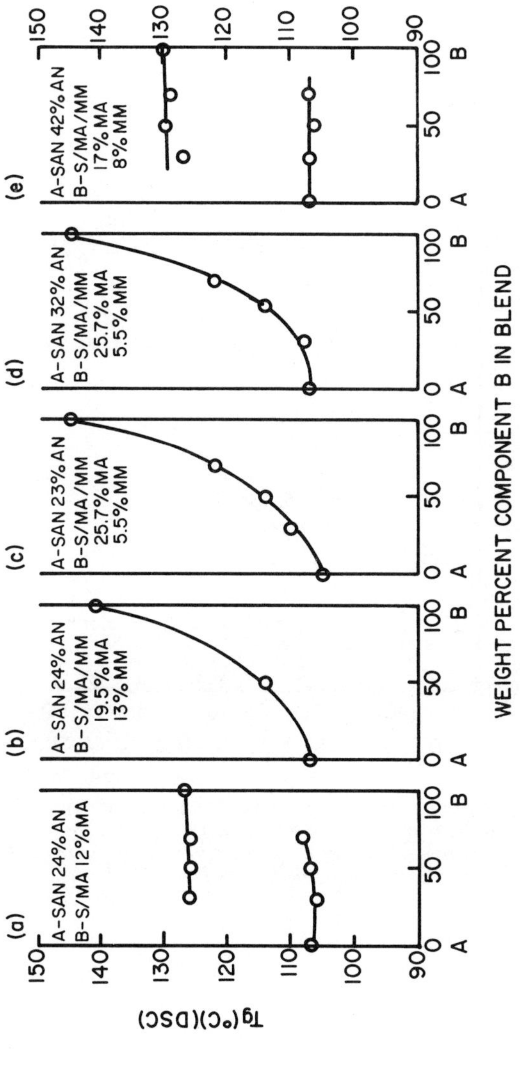

Figure 3. Glass transition temperature–composition plots for various SAN||S/MA/MM blends to illustrate miscibility behavior.

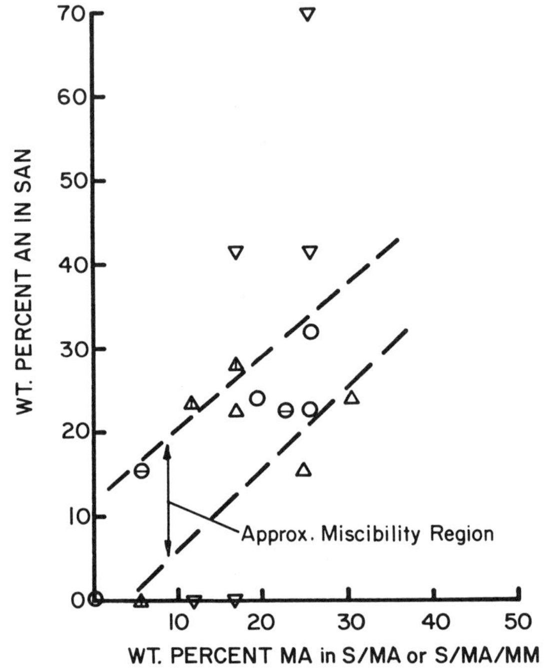

Figure 4. Miscibility diagram for SAN//S/MA/MM blends to show region of miscibility in terms of AN content in the SAN copolymer vs. MA content in the co- and terpolymers. Key: ○, miscible; △, partly miscible; and ▽, immiscible. Slash within the symbols indicates the copolymer S/MA.

Table II. Terpolymer/ABS Blends - Effect of Composition on DTUL

Blend Components			Overall Based on Total Blend					**DTUL
% Terpolymer	% ABS	% α-MS/AN	% MA	*% AN	% IB	% MM	% Rubber	(°C)
100	0	0	22.3	5.1	--	--	14.3	136
60	40	0	13.4	3.1	--	--	21.9	112
50	38	12	11.1	2.5	--	--	20.0	109
100	0	0	28.8	--	5.8	--	17.6	153
85.5	14.5	0	24.7	--	4.9	--	20.8	141
60	31.6	8.4	17.3	--	3.5	--	20.6	118
50	35.7	14.3	14.4	--	2.9	--	20.5	113
48(G)***	52	0	12.2	--	--	--	20	109
48(G)	52	0	12.2	--	--	6.7	20	112
50(G)	39.2	10.8	12.7	--	--	--	12.8	113
50(G)	39.2	10.8	12.7	--	--	2.7	12.8	113
50(G)	39.2	10.8	12.7	--	--	7.0	12.8	115
50(G)	39.2	10.8	12.7	--	--	8.5	12.8	114
50(G)	39.2	10.8	12.7	--	--	10.5	12.8	111
100	0	0	22.3	--	--	5.2	14.1	133
70	30	0	15.6	--	--	3.6	21.8	114
60	40	0	13.4	--	--	3.1	22	110
50	39.2	10.8	11.5	--	--	2.6	20	109

* Overall % AN given includes only terpolymer contribution

** DTUL - 12.7 x 3.2 x 127 mm, compression molded, unannealed specimens, 1.82 MPa

*** (G) indicates glassy terpolymer, all others are rubber-modified terpolymer

consistent with the effects on DTUL of terpolymer, alone, as illustrated in Figures 1 and 2. The effect of varying the terpolymer/ABS ratio on polymer deformation at elevated temperature is shown differently in Figure 5 where the percent retention of room temperature tensile strength (engineering stress) at yield and at fail measured at 100°C is plotted against percent MA present for a particular example of terpolymer and ABS blended in various proportions. Finally, Table III presents a comparison of tensile strength at yield and at fail as a function of temperature from -20°C to 100°C for a typical commercial ABS and for a terpolymer/ABS blend containing 8.3% MA and 19% rubber. Here, the absolute values are less important, in part because the two ABS materials are not comparable (in order to maintain approximate rubber level equivalence), than is the observation of relatively slowly decreasing strength with increasing temperature all the way to 100°C in the case of the terpolymer blend compared to the catastrophic decrease above ca. 60-80°C in the case of this ABS as it softens.

Another aspect of the behavior of polymeric materials in response to heat involves the long-term effects on properties consequent to elevated temperature exposure, particularly in the presence of air. Evaluation of such effects is frequently accomplished by forming test specimens for standard physical testing, exposing these specimens to a set elevated temperature for varying periods of time, and observing the progressive changes (if any) of properties over the time of exposure. Such an experiment was performed on the two materials (18% rubber ABS and 8.3% MA, 19% rubber S/MA terpolymer//ABS blend) for which temperature-dependent tensile data have already been discussed (see Table III). As previously noted, tensile, Izod and DTUL (12.7 x 3.2 x 127 mm) specimens were injection molded and placed in an accurately controlled 90°C circulating air oven (except for the zero aging time specimens which were directly conditioned at 23°C and 50% R.H. for testing). Requisite numbers of specimens were removed from the oven at specified times, cooled and conditioned for testing as previously described. All testing was carried out at 23°C, 50% R.H. (except, of course, DTUL). Figures 6-8 show, respectively, the DTUL, tensile, and notched Izod results as a function of aging time up to 1500 hrs. (ca. 2 months), at which time the test was terminated. (Figure 7 shows only tensile strength at yield (T_y) and elongation at fail (ε_f); tensile strength at fail (T_f) paralleled T_y and is not shown.) As seen in Figure 6, both terpolymer/ABS blend and ABS DTUL's undergo a significant short-time (within one hour at 90°C) increase, which may be ascribed to the frequently observed annealing effect. Thereafter, ABS DTUL appears to remain constant, while the terpolymer/ABS blend DTUL shows a significant additional rise at long times, indicating continuing increase in elevated temperature rigidity with exposure time, perhaps reflecting the greater ΔT from Tg in the terpolymer blend case

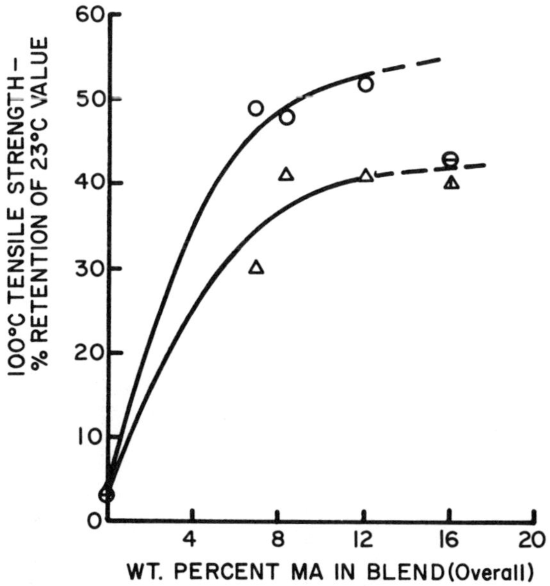

Figure 5. Tensile strength of S/MA terpolymer//ABS blends at 100°C expressed as percent retention of room-temperature value vs. weight percent of maleic anhydride in blend. Key: ○, T_y; and △, T_f. Slash within the symbols indicates that 4% of the sample is a low molecular weight additive.

Table III. *Tensile Strength at Yield and Fail as Function of Temperature

Temp. (°C)	Typical ABS (18% Rubber)		S/MA Terpolymer//ABS Blend (Specific Formulation- 8.3% MA, 19% Rubber)	
	Ty (MPa)	Tf (MPa)	Ty (MPa)	Tf (MPa)
-20	53.2	49.8	48.5	45.9
0	50.6	43.8	40.2	36.0
23(R.T.)	38.7	31.0	30.4	25.8
40	31.9	26.2	25.3	22.8
60	26.4	20.1	21.1	17.5
80	18.5	10.4	17.6	14.4
100	1.1	1.1	14.7	10.5

* Tensile strength given as engineering stress based on original cross-section

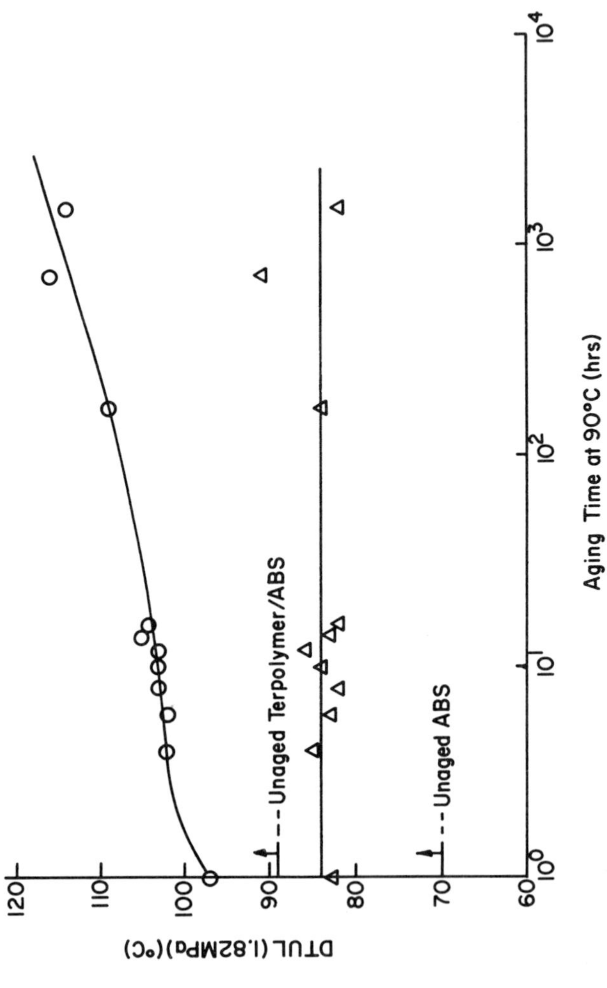

Figure 6. Distortion temperature under load (DTUL) as a function of aging time in oven air at 90°C to compare blend behavior. Key: △, ABS; and ○, S/MA terpolymer/ABS.

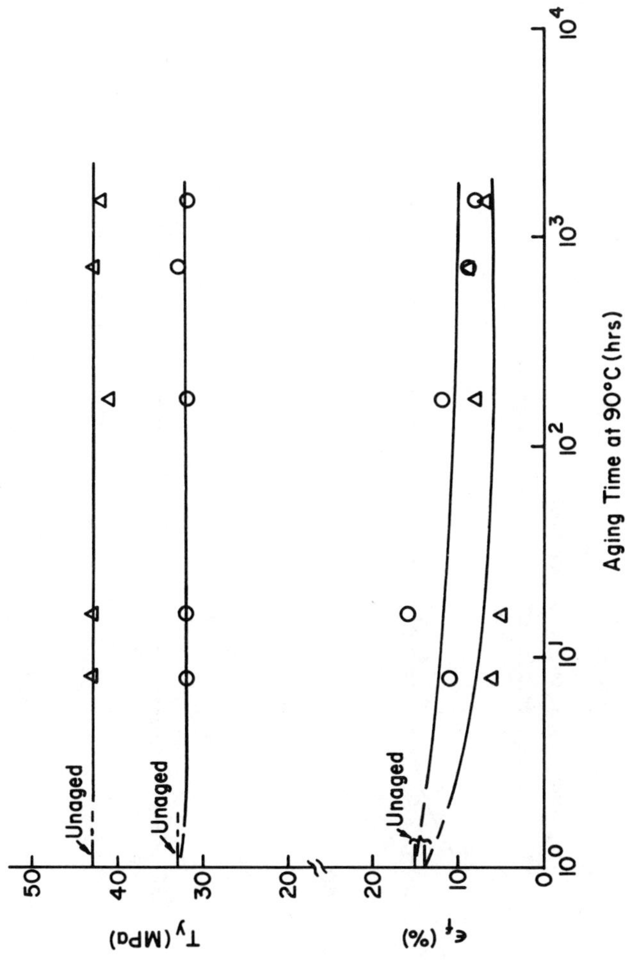

Figure 7. Tensile strength at yield (T_y) and elongation at fail (ϵ_f) as functions of aging time in oven air at 90°C to compare blend behavior. Key: △, *ABS; and* ○, *S/MA terpolymer/ABS.*

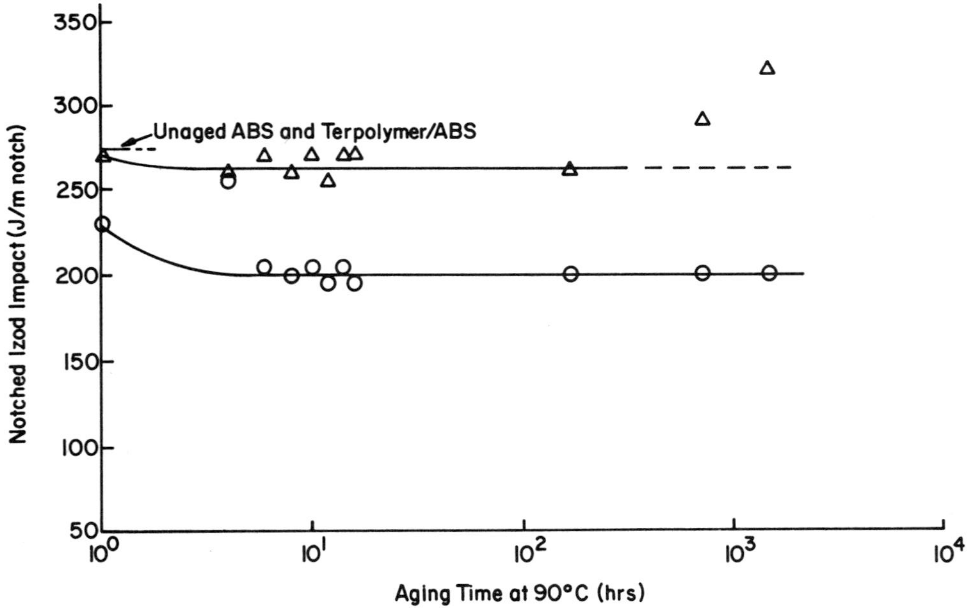

Figure 8. Notched Izod impact as a function of aging time in oven air at 90°C to compare behavior. Key: △, *ABS; and* ○, *S/MA terpolymer/ABS.*

and continuing slow annealing. Over the same exposure time tensile strength (Figure 7) remains essentially unchanged for both materials, and elongation at fail decreases significantly, although less so for the terpolymer/ABS blend than for the ABS sample. (With respect to the tensile strength data, relatively minor differences between the unaged results and the 23°C data in the temperature-dependence study (Table III) are ascribable to the fact that these investigations were carried out separately, with separate moldings and minor modifications to testing protocol between aging and temperature-dependence studies.) Notched Izod impact results as a function of aging time, given in Figure 8, show a significant initial short-time drop in impact strength in the case of the terpolymer/ABS blend, but not in the ABS case. This result is not fully understood at this time. Over longer exposure times both materials appear to reach constant impact strengths, although the ABS shows an inexplicable rise in strength at the longest times (720 and 1500 hrs.). It is clearly to be expected that at some exposure time beyond those covered here both impact strength and elongation at fail must undergo severe decreases for both materials as the toughening rubber particles are degraded. Thus, longer aging time experiments may be expected to shed additional light on the effect of this hostile environment on the blends of interest. The results of such tests may be the subject of future reports.

Summary

A wide compositional range of S/MA/X terpolymers, where X includes acrylonitrile, ethyl acrylate, isobutylene, methyl acrylate, and methyl methacrylate have been studied in terms of the effect of terpolymer composition on resistance to deformation under stress at elevated temperatures. In addition, miscibility of selected terpolymer systems with SAN and physical properties of terpolymer/ABS blends have been defined. In particular the resistance to heat, both elevated temperature rigidity and maintenance of desirable properties in spite of exposure to air and high use temperatures, has been investigated for selected materials. Blending of these terpolymers with ABS provides the opportunity for optimizing a balance of properties with significantly enhanced resistance to the hostile environment represented by high temperature utilization.

Acknowledgment

The authors wish to acknowledge the very considerable contribution to this paper from the work of the late Dr. Y. C. Lee and to thank the Monsanto Company for permission to publish this paper.

Literature Cited

1. Olabisi, O., Robeson, L. M., Shaw, M. T., "Polymer-Polymer Miscibility", Academic Press, N.Y., 1979, Ch.5, pp.215-276.
2. Lee, Y. C., unpublished work.
3. Lee, Y. C., Trementozzi, Q. A., U.S.Patent 4,197,376 (1980)
4. Lee, Y. C., Trementozzi, Q. A., U.S.Patent 4,305,869 (1981)
5. MacKnight, W. J., Karasz, F. E., Fried, J. R., in "Polymer Blends," Paul, D. R., Newman, S., Eds., Academic Press, N. Y., 1978, Vol. 1, Ch. 5, pp. 185-201.
6. Van Krevelen, D. W., Hoftyzer, P. J., "Properties of Polymers", Elsevier Scientific Publishing Company, Amsterdam, 1976, 2nd ed., Ch. 7, pp. 129-159.
7. Molau, G. E., J. Polym. Sci., Polym. Letters, 1965, 3, 1007-1015.

RECEIVED February 4, 1983

6

Polyphenylene Sulfide (PPS) in Harsh Environments

VAN C. VIVES, J. S. DIX and D. G. BRADY

Phillips Petroleum Company, Research and Development, Bartlesville, OK 74004

Hostile environments for plastics generally refer to high temperatures and corrosive chemicals. As high performance plastics take on an increasing role in metal replacement, there is growing emphasis placed on their resistance to these hostile environments. Polyphenylene sulfide is one of these high performance materials whose environmental resistance has been studied extensively. Its long term thermal stability is unique among processable thermoplastic materials and has led to UL ratings of 200-240°C. Its excellent chemical resistance to a wide range of environments has resulted in its consideration and/or use in a number of demanding applications including protective coatings, equipment housings, under-the-hood automotive parts, pump components, valves and oil field hardware. Pertinent environmental testing data on polyphenylene sulfide and its compounds are presented, new applications discussed, and where possible comparisons made to other plastics.

There is a growing list of plastic materials capable of "meeting the challenge" in many applications involving hostile environments. Polyphenylene sulfide (PPS) is one of the leaders on this list and its environmental resistance has been studied extensively. This paper summarizes the performance characteristics of PPS in harsh environments.

A commercial process for polyphenylene sulfide, $(-\langle O \rangle - S -)_x$, was developed in 1967 by the reaction of para-dichlorobenzene and sodium sulfide in a polar solvent (1). This polymer is very versatile. Selected grades can be used for extrusion and for production of a wide range of injection molding compounds. Typical properties of a 40% glass-reinforced compound are given in Table I.

0097–6156/83/0229–0065$06.00/0
© 1983 American Chemical Society

Table I. Properties Of 40% Glass-reinforced PPS*

Flexural Strength	156 MPa	(22,700 psi)
Tensile Strength	121 MPa	(17,500 psi)
Flexural Modulus	1.17×10^4 MPa	(1.7×10^6 psi)
Compressive Strength	145 MPa	(21,000 psi)
Izod Strength		
Notched	58 J/m	(1.1 ft-lbs/in)
Unnotched	175 J/m	(3.3 ft-lbs/in)

*Ryton R-4 Polyphenylene Sulfide

Chemical Resistance

The unique aromatic sulfide backbone of the polyphenylene sulfide molecule is, undoubtedly, responsible for its heat and chemical resistance. Resistance to various organic and inorganic compounds and reagents, compared to other polymers, is excellent. Short-term exposure tests were made at 93C and 24 hours on injection molded bars of PPS containing no fillers or reinforcing agents. The tests included 127 compounds and reagents, and only 16 caused more than 20% reduction in tensile strength. The list of those causing significant deterioration of tensile strength includes aqua regia, chlorosulfonic acid, 50% chromic acid, 96% sulfuric acid, some chlorinated compounds, and some amines. Excellent resistance to a wide variety of other chemical environments was demonstrated (2).

Results of some of these short-term tests are shown in Table II. A comparison is given between PPS and five other plastics: nylon (Zytel 101), polycarbonate (Lexan 141), polysulfone (Bakelite Polysulfone), polyphenylene oxide (Noryl), and polyetherimide (Ultem 2300). The data presented are based upon retention of tensile strength for all plastics except the Ultem 2300, which is based upon retention of flexural strength. Unsuccessful attempts were made to injection mold ASTM Type IV tensile bars out of the Ultem compound, but flexural strength bars could be made. Experience has shown that chemical resistance tests monitored by flexural strength retention are comparable to those monitored by retention of tensile strength.

Polyphenylene sulfide has very good resistance to acids and bases. Formic acid decreases the tensile strength by 25%. Resistance to alcohols and glycols is excellent. Of all the classes of organic compounds tested, amines cause the most deterioration in tensile strength.

Table II. Chemical Resistance At 93°C For 24 Hours[a]

	Percent Retention of Tensile Strength[b]					
	PPS	Nylon 6-6	Polycarbonate	Polysulfone	Modified PPO	Polyetherimide
Acids						
10% Nitric	96	0	100	100	100	93
37% HCl	100	0	0	100	100	90
10% Acetic	100	30	37	100	100	---
Glacial acetic	98	0	67	91	78	---
88% Formic	75	0	38	79	99	---
30% Sulfuric	100	0	100	100	100	---
Bases						
15% NaOH	100	69	98	100	100	---
30% NaOH	100	89	7	100	100	69
28% NH_4OH	100	85	0	100	100	0
Alcohols						
Butyl	100	87	94	100	84	100
2-Aminoethanol	100	93	0	100	100	72
Cyclohexanol	100	84	0	95	27	---
Ethylene glycol	100	96	100	100	100	98
Aldehydes & Ketones						
Benzaldehyde	84	98	0	0	0	---
Cyclohexanone	99	87	0	0	0	0
Methyl Ethyl Ketone	100	87	0	0	0	56
Amines						
Butylamine	49	91	0	0	0	---
Ethylenediamine	65	78	0	0	51	0
Aniline	96	85	0	0	0	0
Pyridine	93	74	0	0	0	---
Morpholine	80	93	0	0	0	---
Esters						
Ethyl Acetate	100	89	0	0	0	86
Amyl Acetate	100	88	46	0	0	---
Butyl Phthalate	100	90	46	63	19	---

Table II--Continued on next page

Table II, Continued
Chemical Resistance At 93°C For 24 Hours[a]

	Percent Retention of Tensile Strength[b]					
	PPS	Nylon 6-6	Polycar-bonate	Poly sulfone	Modified PPO	Poly-ether-imide
Ethers						
Butyl Ether	100	200	61	100	0	98
Tetrahydrofuran	76	87	0	0	0	---
p-Dioxane	88	96	0	0	0	---
Halogen Compounds						
Chloroform	87	57	0	0	0	0
Ethylene Chloride	72	65	0	0	0	0
Chlorobenzene	100	73	0	0	0	0
Benzyl Chloride	100	80	0	0	0	---
Hydrocarbons						
Heptane	91	84	100	100	36	---
Cyclohexane	100	90	75	99	0	---
Toluene	98	76	0	0	0	61
Xylene	100	91	0	0	0	81
Miscellaneous						
Acetonitrile	96	93	25	0	69	---
Benzonitrile	100	88	0	0	0	---
Nitrobenzene	100	100	0	0	0	0
Phenol	100	0	0	0	0	---
N,N-dimethyl-formamide	100	95	0	0	---	---

a. The polymers tested were: Ryton Polyphenylene Sulfide, no filler or reinforcing agent, Phillips Petroleum Co.; Zytel 101 Nylon, E. I. duPont deNemours & Co., Inc.; Lexan 141 Polycarbonate, General Electric Co.; Bakelite Polysulfone, Union Carbide Corp.; Noryl Polyphenylene Oxide, General Electric Co.; Ultem 2300 Polyetherimide, General Electric Co.

b. The values are for retention of tensile strength except for the polyetherimide, which are for retention of flexural strength.

In Figure 1 the percentage of reagents which caused less than 25% reduction in tensile strength have been plotted for PPS and five other plastics, including a phenolic thermoset. PPS shows better resistance than even the phenolic resin, which is known for its chemical resistance.

Long-term tests were made at 93°C on 40% glass-reinforced PPS. Even with extended exposure times of one year, which cause catastrophic failure in many thermoplastics, the resistance of PPS is very good. (Table III) It is common in long-term testing to regard 50% or more retention of property as a "pass" test. Specimens of PPS (glass reinforced) having 50% retention of tensile strength still possess greater than 55 MPa (8,000 psi) of strength, which is adequate for many applications.

Table III. Long-Term Chemical Resistance of 40% Glass-Reinforced PPS at 93°C

	Per Cent Retention of Tensile Strength	
	3 Months	12 Months
37% Hydrochloric Acid	34	29
30% Sulfuric Acid	89	61
10% Nitric Acid	0	0
85% Phosphoric Acid	99	89
30% Sodium Hydroxide	89	63
n-Butyl Alcohol	92	80
Cyclohexanol	91	86
Phenol	92	63
Butylamine	46	--
Aniline	86	42
Benzaldehyde	47	42
Methyl Ethyl Ketone	100	61
N-Methylpyrrolidone	92	80
Chloroform	77	43
Carbon Tetrachloride	48	25
Ethyl Acetate	88	58
Butyl Ether	89	79
p-Dioxane	96	59
Toluene	70	41
Nitrobenzene	63	31

Aqueous nitric and hydrochloric acids caused considerable reduction in tensile strength. Inorganic compounds are usually tested as aqueous solutions and it is very often the more dilute solutions which have the most deleterious effect upon glass-reinforced PPS. This is believed to be caused by the action of water, not the reagent, on the glass-resin interface. More will be said about this in the section on automotive uses.

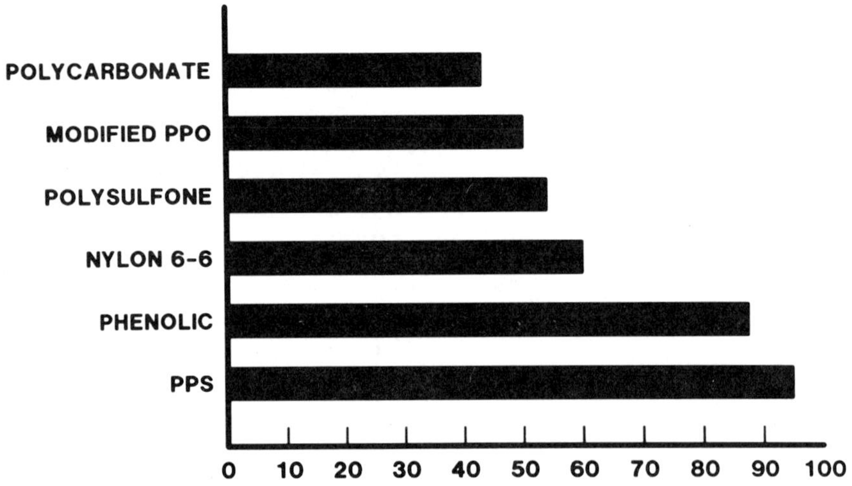

Figure 1. Chemical resistance at 93°C for 24 h.

Butyalmine, aniline, benzaldehyde, chloroform, carbon tetrachloride, toluene, and nitrobenzene caused more than 50% reduction in tensile strength after exposure for one year. However, only butylamine, benzaldehyde, and carbon tetrachloride reduced tensile by more than 50% at three months of exposure.

Summarizing chemical resistance data, over a broad range of chemicals, PPS is second only to fluorocarbon polymers such as polytetrafluoroethylene (3).

Resistance To Automotive Fluids

There is increasing interest in the application of reinforced plastics for metal replacement in the automotive industry to reduce weight and cost. In some applications, reinforced plastics also offer performance advantages over metal. Although the automotive industry is greatly depressed at the present time, a recent report indicates an 11.6% annual growth rate for reinforced plastics in cars and trucks in the next five years (8).

Automotive underhood environments are often antagonistic to plastics. Extreme temperatures coupled with exposure to oil, fuel, and coolants will permit only the most resistant of plastic compositions to be used. The heat and chemical resistance of PPS makes it a prime candidate for such cases.

Gasoline. Polyphenylene sulfide has very good resistance to gasoline. Unfilled PPS exposed to gasoline for 24 hours at 93°C exhibited 100% retention of tensile strength. There was no significant change in tensile strength when 40% glass-reinforced PPS was exposed to gasoline containing 15% methanol or ethanol for twelve weeks at 121°C. (Figure 2) Long-term exposure to gasoline containing 8% of the octane booster, methyl tert-butyl ether (MTBE), caused no decrease in tensile strength. (Figure 3) Rynite 545, a glass-reinforced polyethylene terephthalate, lost over 90% of its tensile strength after only two weeks in gasoline/methanol at 121°C exposure. Weight and thickness changes for glass-reinforced PPS and polyetherimide (Ultem 2300) are shown in Table IV. Changes are greater for gasoline/methanol than for gasoline or gasoline/ethanol for both plastic compounds. Ultem 2300 is affected more than PPS by the 121°C exposure.

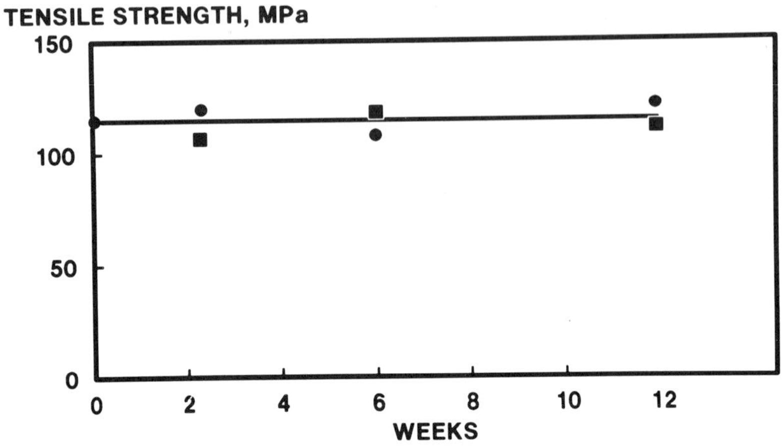

Figure 2. Exposure of 40% glass-reinforced PPS to gasoline/15% alcohol mixtures at 121°C. Key to alcohol: ■, methanol; and ●, ethanol.

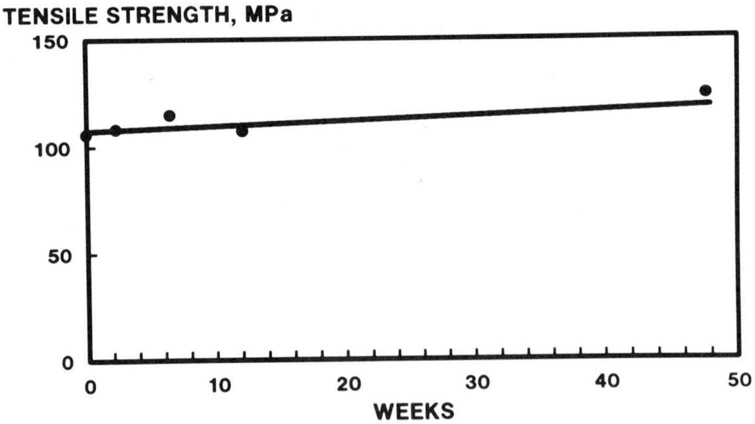

Figure 3. Exposure of 40% glass-reinforced PPS to a gasoline/8% MTBE mixture at 121°C.

Table IV. Weight and Thickness Changes of 40% Glass-reinforced PPS andUltem 2300 Polyetherimide on Exposure to Gasoline andGasoline/Alcohol at 121°C

	% Weight Change		% Thickness Change	
	PPS	Ultem	PPS	Ultem
Gasoline				
1 Week	+ 0.02	-0.10	+ 0.40	+ 0.69
2 Weeks	+ 0.08	-0.09	+ 0.05	+ 0.72
4 Weeks	+ 0.13	-0.13	+ 0.37	+ 1.07
Gasoline/15% Methanol				
1 Week	+ 0.69	+ 2.08	+ 1.13	+ 5.47
2 Weeks	+ 0.65	+ 2.99	+ 0.70	+ 6.26
4 Weeks	+ 0.59	+ 4.07	+ 0.97	+ 6.42
Gasoline/15% Ethanol				
1 Week	+ 0.36	+ 0.37	+ 0.49	+ 2.76
2 Weeks	+ 0.38	+ 0.66	+ 0.83	+ 2.58
4 Weeks	+ 0.45	+ 0.84	+ 0.81	+ 3.53

Oil. Other underhood fluids include oils, such as motor oil, brake, and transmission fluids. Exposure of glass-reinforced PPS to motor oil for four weeks at 180°C (356°F), a severe temperature for most plastics, caused very little decrease in strength. (Table V) Due to formation of various oxygenated organics during conventional engine service, used engine oil might be expected to be more harsh on plastic materials than fresh oil. The Cummins Engine Company conducted a long-term study of exposure of glass-reinforced PPS to used engine oil at 150°C (302°F). (Figure 4) (5) Nylon was included in the test since it is also a candidate material for applications involving exposure to used engine oil. After a small decrease in flexural strength the PPS specimens stabilize, whereas the nylon continues to decrease slightly after one year.

Table V. Resistance Of 40% Glass-Reinforced PPS To SAE 30 Motor Oil at 180°C/4 Weeks

	% Retained
Flexural Strength	97
Tensile Strength	87
Notched Izod Impact	98

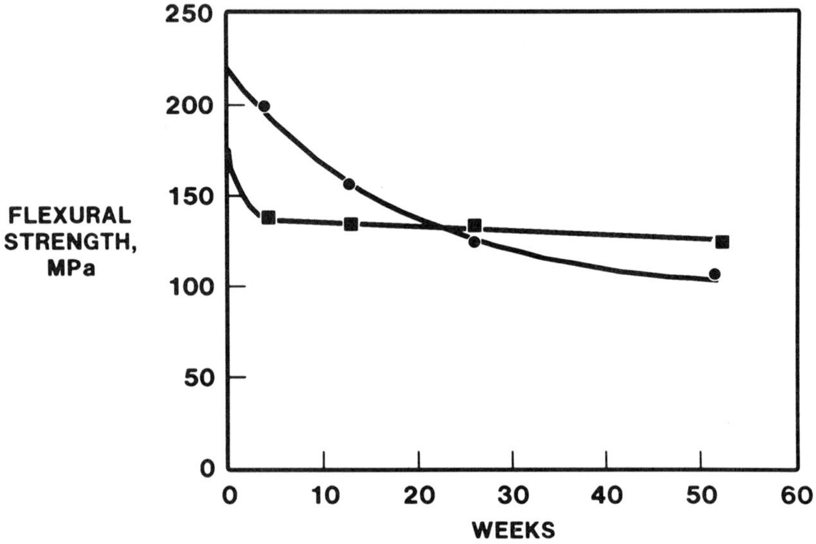

Figure 4. Exposure of glass-reinforced PPS (■) and nylon (●) to used engine oil at 150°C.

Reinforced PPS exhibited excellent resistance to brake and transmission fluids at 149°C (300F) for one year (Figure 5).

Coolants. Underhood fluids also include water and antifreeze solutions. Physical properties of glass-reinforced plastics of all types deteriorate, sometimes drastically, upon exposure to water at elevated temperatures. Bair and Miner of Bell Laboratories have recently studied the hydrolytic stability of PPS and glass-reinforced PPS using a calorimetric method (6). The results which they obtained agree well with our data. (Figure 6)

The physical properties of glass-reinforced plastics of all types generally deteriorate upon exposure to water at elevated temperatures (7). Even if the resin itself is hydrolytically stable, the water penetrates the molded part by a wicking action and attacks the glass-resin interface causing delamination. Reinforced PPS is no exception, but the rapid initial decrease in mechanical properties soon stabilizes and no extensive further deterioration is observed. Long-term tests with water and engine coolant were made by Cummins Engine Company. (Figure 7) After about 10 to 20 weeks of exposure, the flexural strength stabilizes. This is very important, because molded parts can then be engineered for the final strength, which is still 75-80 MPa (11,000 psi). Compared to phenolic resin, considered to be very resistant, PPS stabilizes at a higher strength, while the phenolic material continues to lose strength even after one year. (Figure 8)

Thermal Stability

PPS Compounds possess exceptional short-term and long-term thermal stability. Thus, the tensile strength of 40% glass-reinforced PPS aged at 230°C (446°F), after an initial loss of 20%, gradually decreases to about 50% of initial strength after 10,000 hours. Based upon laboratory studies such as this, the family of PPS compounds has been given the highest UL temperature indexes of any thermoplastic. This index is an indication of the service temperature that a material can tolerate. As shown in Table VI, temperature indexes range from 200°C to 240°C depending upon filler content, thickness and end-use.

Table VI. UL Temperature Index Of PPS Compounds

Compound	Min. Thickness (mm)	Temperature Index, °C Elect.	Temperature Index, °C Mech. with Impact
Glass-Reinforced	.71	200	200
	1.47	220	200
Glass-Reinforced, Mineral Filled[a]	.71	200	200
	1.47	220	220
Glass-Reinforced, Mineral Filled[b]	.71	220	200
	1.47	240	220

[a] Ryton R-8 [b]Ryton R-10 7006A

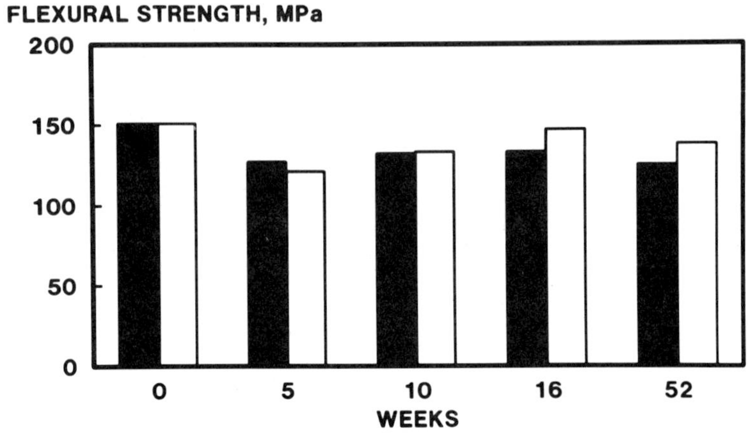

Figure 5. Exposure of 40% glass-reinforced PPS to dot-3 brake fluid (closed bar) and type-f transmission fluids (open bar) at 149°C.

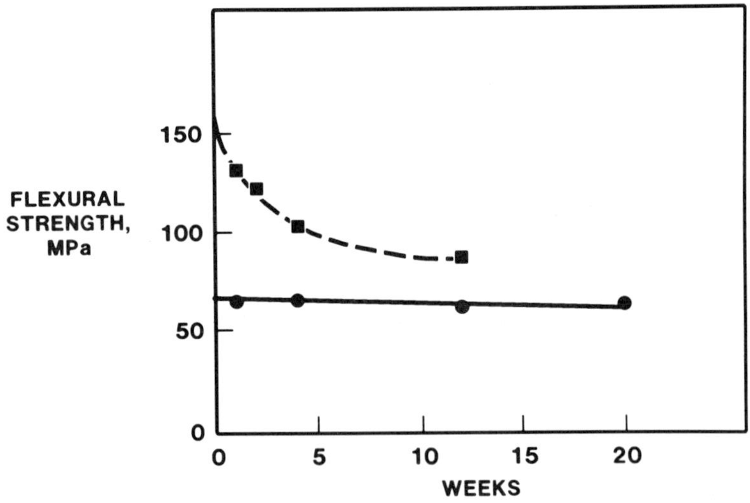

Figure 6. Exposure of unfilled (■) and 40% glass-reinforced (●) PPS to water at 95°C.

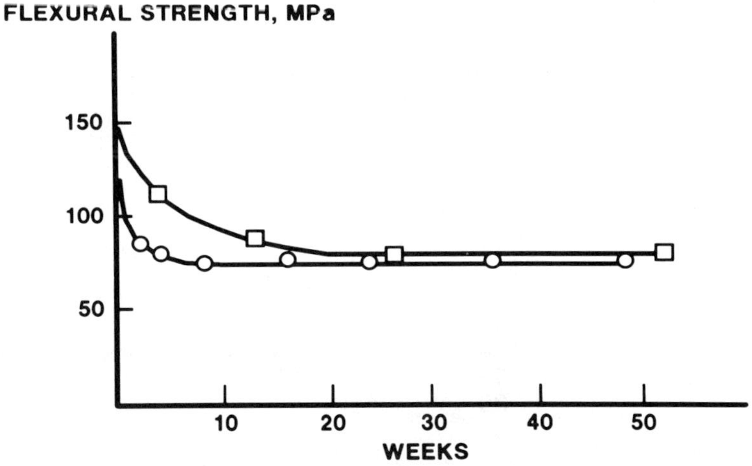

Figure 7. Exposure of 50% glass-reinforced PPS to water at 120°C (○) and engine coolant at 125°C (□).

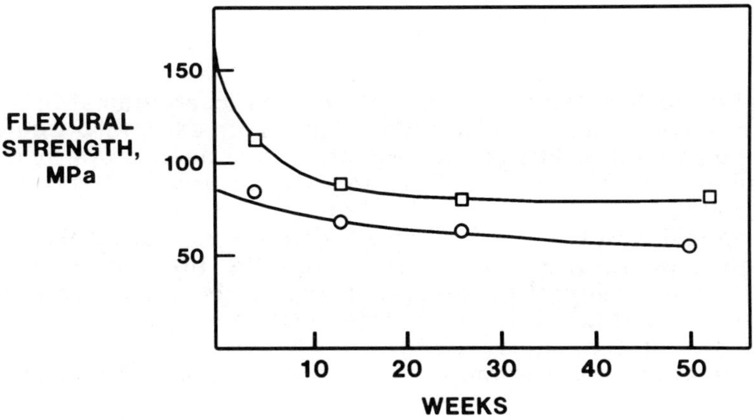

Figure 8. Exposure of 40% glass-reinforced PPS (□) and a phenolic (○) to engine coolants at 125°C.

Fibers can be produced from a specific grade of PPS polymer. This polymer has the same outstanding heat resistance possessed by the grade of material used in injection molding compounds. As shown in Table VII, continuous filament PPS yarn has been exposed at 200°C for 60 days with only a moderate decrease in strength.

Table VII. Thermal Stability Of PPS Fiber

Exposure Time (Days)	Retained Strength Compared To No Exposure, %		
	Tenacity[a]	Knot Tenacity[b]	Elongation[c]
1	146	145	100
7	125	127	95
13	111	127	86
20	107	132	77
27	111	122	77
33	114	104	77
40	93	95	68
47	89	91	59
54	82	95	54
61	86	91	50

260/34 Yarn exposed at 200°C

[a] Initial value 2.8 gpd
[b] Initial value 2.2 gpd
[c] Initial value 22%

Heat deflection temperature is often used as an indication of heat resistance for short time periods, such as in assembly operations involving soldering. PPS compounds that have been properly molded and heat treated have heat deflection temperatures (ASTM D648) over 260°C (500°F).

Creep resistance is also an important property for applications at elevated temperatures. As shown in Figure 9, both glass-reinforced and glass-and-mineral-containing compounds have good creep resistance as measured by the change in creep modulus. In a study of compressive creep, glass-reinforced PPS had the highest creep resistance of the four thermoplastic materials studied (Table VIII) (8).

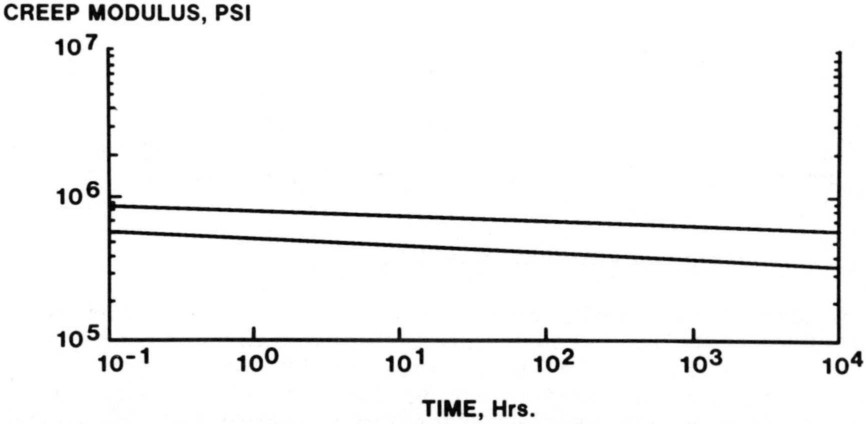

Figure 9. Creep modulus of glass- and mineral-filled PPS compounds at 3000 psi (top) and 40% glass-reinforced PPS compounds at 5000 psi (bottom) all at 121°C.

TABLE VIII

MATERIAL RANKING BY CREEP RESISTANCE

AND HEAT DEFLECTION TEMPERATURES

Material	Material Ranking, Creep[a] 93°C 20.7MPa	93°C 69MPa	149°C 20.7MPa	149°C 69MPa	Overall Average	HDT,°C ASTM D648
Phenolic[b]	1	1	1	1	1	>260
Polyphenylene Sulfide[c]	2	2	2	2	2	>260
Polybutylene Terephthalate[d]	3	4	3	3	3	218
Thermoplastic Polyester[e]	4	3	4	3	4	218
Nylon 66[f]	5	4	5	3	5	49

a. 20.7 MPa = 3000 psi; 69 MPa = 10,000 psi
b. Glass/mineral filled
c. 40% Glass fiber
d. 30% Glass fiber
e. Glass/mineral filled PET
f. 30% Glass fiber

Property Maintenance At Elevated Temperature

A second aspect of thermal resistance involves retention of properties at elevated temperatures. Determination of properties of 40% glass-reinforced molding grade of PPS over temperatures ranging from room temperature to 205°C (400°F) shows classical deterioration above the glass transition temperature. Fortunately, because of crystallinity effects and good compatability with the glass reinforcement the loss in strength is gradual and even at temperatures of 205°C considerable strength integrity is retained. In general, with exception to impact strength which increases as temperature increases, 40% glass-reinforced moldings retain about 80% of their original strength at 100°C (212°F), 60% at 160°C (320°F) and 40% at 205°C (9).

Weatherability

Outside exposure to the elements can certainly take its toll on plastic materials, particularly those not properly protected by stabilizers. Unstabilized tensile bars of 40% glass-filled PPS were aged in an Atlas Weatherometer for 10,000 hours. Moderate stability was observed as shown in Table IX. The loss in tensile strength was reduced from 37% to 9% by adding 1% Tinuvin 327 stabilizer.

Table IX. Weather-ometer Tests On 40% Glass-reinforced PPS

Hours Exposed	Tensile Strength, MPa	
	No Stabilizer	1% Tinuvin 327
0	115	113
2,000	105	109
6,000	106	104
8,000	98.9	97.8
10,000	72.7	103

Flammability

Although PPS, like most organic materials, will burn when exposed to an external flame, its aromatic chemical structure and its innate ability to char when exposed to an external flame, provide inherent flame resistance. In its unfilled form it exhibits a high oxygen index of 44. Addition of a glass and mineral filler raises the index to ca. 55. PPS provides a low radiant panel flame spread index (ASTM E-162) and according to standard UL (UL94) laboratory flammability tests is classified as V-O/5V (i.e., self extinguishing, non-dripping). Unlike other aromatic polymers such as polystyrene and inherently flame resistant structures such as polyvinyl chloride, PPS does not produce large amounts of smoke in either the smoldering or flaming states (NBS Test Chamber). PPS has an auto-ignition temperature of 540°C (1004°F). Comparative oxygen indexes and smoke density data are provided in Tables X and XI (10, 11).

Table X. Oxygen Index Values For Plastics

Plastic	Oxygen Index
Polyolefin	18
SAN	19
ABS	19
Nylon	24
Polycarbonate	24
PPO	25
Polysulfone	30
Polyimide	37
PPS	44-55
PVC	45
PTFE	95

Table XI. Smoke Density.
National Bureau Of Standards Test

Material, (thickness)	Maximum Specific Optical Density, Dm A	B	Obscur. Time (D_s = 16)[f], min. A	B
Red Oak (6.4 mm)	393	75	4.1	80
Polystyrene (6.4 mm) (6.4 mm)[a]	395	780	4.0	0.63
Polyvinyl Chloride (3.2 mm)	270	525	24.0	0.50
Nylon 6/6 (3.2 mm)[b]	487	41	3.6	9.20
Polysulfone (6.4 mm)[c]	111	370	12.6	1.90
Polycarbonate (3.2 mm)	21	324	27.0	2.00
Phenolic (3.2 mm)[d]	110	50	5.0	5.80
Polyphenylene Sulfide (6.4 mm)[e]	25	232	15.5	3.20

A = Smoldering B. = Flaming

a. UCC SMD -3500
b. 40% glass filled
c. P-1700
e. Ryton PPS R-4, 40% glass filled
f. Time required for specific optical density(Ds) to reach value of 16.

Radiation Resistance

Relatively high exposures of gamma radiation have no significant effect on the mechanical properties of glass-reinforced and glass- and mineral-containing PPS compounds. (Table XII) As shown in Table XIII, neutron radiation has only a slight effect on these compounds.

Table XII. Effect Of Gamma Radiation On Mechanical Properties Of PPS Compounds[a]

Compound	Retained Strength Compared to No Exposure, % Flex Str	Ten Str	Flex Mod
40% Glass-reinforced[b]	93	100	102
Glass-reinforced, mineral-filled[c]	94	101	102
Glass-reinforced, mineral-filled[d]	99	98	102

a. Radiation dose of 3×10^8 rad
b. Initial values: FS, 184 MPa; TS, 148 MPa; FM, 1.27×10^4 MPa
c. Ryton R-8, initial values: FS, 111 MPa; TS, 83 MPa; FM, 1.22×10^4 MPa.
d. Ryton R-10 7006A, initial values: FS, 144 MPa; TS, 95 MPa; FM 1.49×10^4 MPa

Table XIII. Effect Of Neutron Radiation On Mechanical Properties Of PPS Compounds[a]

Compound	Retained Strength Compared to No Exposure, % Flex Str	Ten Str	Flex Mod
40% Glass-reinforced[b]	86	90	102
Glass-reinforced, mineral-filled[c]	88	90	100
Glass-reinforced, mineral-filled[d]	95	87	96

a Total thermal neutron and gamma dose of $4x10^8$ rad.
b Initial values as given in Table XII.
c Ryton R-8, initial values as given in Table XII.
d Ryton R-10 7006A, initial values as given in Table XII.

Applications

The outstanding thermal stability and chemical resistance of PPS compounds have led to their use in applications ranging from food warmers to nuclear installations. Some of the harsh environments which these compounds are successfully withstanding are shown in the following applications:

Integrated circuit device burn-in test sockets. These chip carriers must survive high continuous test oven temperatures.

Oxygen sensor connector insulators. These parts must withstand the high temperatures of automotive engines.

Oil well sucker rod guides. These components must not only tolerate the extreme temperatures of deep oil wells, but must remain unaffected by the hot oil and corrosive liquids.

Chemical pumps. Pump housings and impellers of PPS compounds withstand corrosive liquids that would destroy metal components.

Carburetors. A PPS compound has replaced aluminum in a carburetor for a small gas engine.

pH meters. In one case, pH meters with PPS components have served over three years without failure in a corrosive chemical application where stainless steel meters typically failed in two months.

Flue stack filter bags. Fabric made of PPS fiber has shown superior resistance to the hot sulfur dioxide gases of coal-fired boilers, thus providing environmental protection.

Electrical connectors. In some applications, electrical components of PPS compounds undergo assembly operations in which they are exposed for brief times to solder baths well above 300°C.
Thermostat housing. These components must survive the heat and chemical attack of automotive coolants for prolonged service periods.
Diesel engine piston cooling nozzles. Before acceptance, these components had to withstand several thousand hours of continuous exposure to the hot lubricating oil of heavy duty diesel engines.
Lighting reflectors. In this application, a high temperature plastic that meets reflectance design requirements and that has long term resistance to ca 240°C is required. Examples include floodlamps and halogen lamps.
Capacitor components. In some capacitor applications, long term resistance to polar organic solvents is required.
Automotive. PPS is often the choice due to thermal stability and chemical resistance. Its compounds are used in emission control, fuel, ignition, cooling, brake and electrical systems.
Industrial valves. A unique ball valve made from a PPS composition has gained wide recognition in the paper-making industry for its effectiveness in the containment of corrosive liquids.
Coatings. The thermal and chemical resistance of polyphenylene sulfide has led to a wide variety of applications as a coating material. Examples are cookware with heat-resistant, easy-release coatings containing PPS, valves and fittings coated with PPS for protection against corrosive chemicals, and electrical coils coated with PPS as a temperature-resistant insulation.
Electrical Bobbins. PPS compounds are used in these components of solenoids and transformers where resistance to temperatures of up to 180°C is required.
Fuel Filters. PPS compounds because of their chemical resistance and temperature stability are suitable for use in this application where service temperatures can range from -40°C to 200°C.
Microwave oven applications. A special grade of PPS compound is produced for use in microwave ovenware, where its chemical resistance and thermal stability combined with low absorption of the microwave energy make it attractive for this application. PPS compounds are also used in components of microwave ovens.

Literature Cited

1. Edmonds, J. T., Jr.; Hill, H. W. Jr. U.S. Patent 3354129, 1967.
2. Brady, D. G.; Hill, H. W., Jr. Modern Plastics, 1974, 51, 60.
3. Plastics World 1977, 35, 30.
4. Plastics Technology 1982, Oct., 101.
5. Miller, J. R.; Dix, J. S.; Vives, V. C. Proc. Automotive Plastics Durability Conf. SAE 1981, 75-78.
6. Plastics Compounding 1982, July/August, 49.
7. Wake, W. C. in "Fillers for Plastics", Wake, W. C., Ed.; Butterworth: London, 1971, Chap. 1.

8. Moska, J. C. Reg. Tech. Conf. Soc. Plast. Eng. 1980, 127-133.
9. Hill, H. W., Jr.; Brady, D. G. in "Kirk-Othmer Encyclopedia of Chemical Technology"; John Wiley: New York, 1982, Vol. XVIII, p. 793.
10. Hilado, C. S. "Oxygen Index of Materials", Fire and Flammability Series, Technomic: Westport, CT. 1973, Vol. IV.
11. Hilado, C. S. "Flammability Handbook for Plastics", Technomic: Westport, CT., 1974, p. 60.

RECEIVED February 22, 1983

7

Advances in High-Performance Polymers

RAYMOND B. SEYMOUR

University of Southern Mississippi, Department of Polymer Science, Hattiesburg, MS 39406-0076

While polymers with moderately high performance qualities, such as cellulose in wood and cotton have been known for years, these linear polymers could not be extruded or injection molded. However, they were able to withstand the effects of moderately hostile environments. The classic reinforced thermosets could be compression molded and could also be classified as high performance plastics. In contrast, the general purpose thermoplastics, introduced in the 1930's, were readily extruded and injection molded but were not useful at boiling water temperatures.

Fortunately, the deficiencies of both the classic thermosets and general purpose thermoplastics have been overcome by the commercialization of a series of engineering plastics including polyacetals, polyamides, polycarbonate, polyphenylene oxide, polyaryl esters, polyaryl sulfones, polyphenylene sulfide, polyether ether ketones and polyimides. Many improvements in performance and processing of these new polymers may be anticipated through copolymerization, blending and the use of reinforcements.

Polymers with moderately high performance qualities, such as cellulose and cotton and wood have been known for centuries. These linear polymers could not be extruded or injection molded but they did withstand the effects of moderately high hostile environments.

The classic thermosets, such as ebonite, phenolics, ureas, melamines and polyesters as well as the epoxies, when reinforced with fiberglass or graphite fibers could also be classified as high performance thermosets. While these plastics cannot be readily extruded or injection molded, they are resistant to the effects of moderately high hostile environments. Phenolic resin mortars have been used for over a half century for the construction of chemical resistant vessels and for joining brick and tile used as linings in hot acid environments.(1)

0097-6156/83/0229-0087$06.00/0
© 1983 American Chemical Society

The general purpose thermoplastics introduced in the 1930's and 1940's were readily fabricated by extrusion and injection molding techniques. Thermoplastics, such as polyvinyl chloride, polystyrene and polyethylene were resistant to mineral acids at temperatures up to 60°C but they could not be used at higher temperatures, such as that of boiling water. These large volume plastics had other characteristic deficiencies which were overcome by the use of additives.

The brittleness of polyvinyl chloride and polystyrene was decreased by blending with plasticizers or impact modifying polymers. The flammability of polystyrene and polyolefins was decreased by the addition of flame retardants and the instability of polyvinyl chloride and polypropylene was reduced by the addition of stabilizers.(2) The strength and heat resistance of all of the general purpose plastics were improved by reinforcing with fiberglass or graphite fibers.

The heat resistance of polyvinyl chloride was also improved by post chlorination, the ductility of polyethylene was improved by increasing the molecular weight and the usefulness of polystyrene was increased by copolymerization with acrylonitrile. Nevertheless, in spite of improvement in performance, resulting from these modifications, most general purpose plastics were under engineered for many applications.

Blends of copolymers of styrene and acrylonitrile and butadiene and acrylonitrile called ABS plastics which are more ductile than polystyrene, are now used at an annual rate of almost 500 thousand tons. Terpolymers of styrene, acrylonitrile and maleic anhydride (Cadon) have heat deflection points above 100°C.(3) While the physical properties of both ABS and the maleic anhydride terpolymers are superior to polystyrene, the improvements are not sufficient to classify them as high performance plastics.

Nevertheless, these medium performance plastics can be used in many applications which are unsuitable for general purpose plastics. More important, these plastics, which are also moderately expensive, may be upgraded by changes in formulation. For example, ABS which, with an annual sale of 450 thousand tons, is the largest selling engineering plastics, can be upgraded by replacing the acrylonitrile-butadiene copolymer elastomer (NBR) with ethylene-propylene copolymer elastomer (EDPM).

The history of early developments in high performance polymers has been recorded by several pioneers in polymer science.(4) These polymers are discussed in this article under specific headings.

Polyacetals (POM)

Polyoxymethylene, which precipitates spontaneously from uninhibited aqueous solutions of formaldehyde, was isolated by

Butlerov in 1859 and investigated by Nobel laureate Staudinger in the 1920's. Polyoxymethylene stabilized by acetylation of the hydroxyl end groups (capping) was introduced commercially by DuPont under the trade name of Delrin in the late 1950's.[(5)]

The structure of the repeating unit in POM is

$$\left[CH_2-O \right]$$

Stable acetal copolymers of formaldehyde, called Hostoform and Celcon were also produced commercially by the cationic copolymerization of formaldehyde with ethylene oxide.[(6)]

The structure of the repeating unit in the copolymer is

$$\left[CH_2-O-CH_2-CH_2-O \right]$$

These high performance plastics are now being produced at an annual rate of 45 thousand tons. It is anticipated that the annual production in the late 1980's will be 80 thousand tons. A typical glass filled (25%) acetal polymer will have the following properties:

tensile strength (psi)	18,500
elongation (%)	3
flexural strength (psi)	28,000
flexural modulus (psi)	1,100,000
izod impact (ft lbs/in of notch)	1.8
Rockwell hardness	M79
melting point (°C)	175
deflection temperature (°C) @ 264 psi	150

The selling price of polyacetal and other engineering plastics is about one-half that of cast metals. Polyacetals have been approved by the Food and Drug Administration for contact with foods. Some of the uses for molded polyacetals are as follows: valves, faucets, bearings, appliance parts, springs, automobile window brackets, hose clamps, hinges, gears, shower heads, pipe fittings, video cassettes, tea kettles, chains, flush toilet float arms, pasta machines, desk top staplers and air gun parts.

These unique polymers are resistant to many solvents, aqueous salt and alkaline solutions and weak acids at a pH greater than 4.5. More complete data on physical properties and resistance to hostile environments of these and other polymers are available.[(7)]

Polyamides

Nylon-66, which was the classic high performance thermoplastic was produced by Carothers in the 1930's by heating the purified salt produced by the reaction of adipic acid and hexamethylenediamine.[(8)] Over 125 thousand tons of these polyamides are used annually as high performance plastics in the U.S. It is anticipated that this use will exceed 180 thousand tons annually in the late 1980's.

Glass filled Nylon-66 with a heat deflection temperature of 250°C (9) and a transparent nylon-66 (10) are available commercially. Molded nylon-66 is being used for lawn mower blades, bicycle wheels, tractor hood extensions, skis for snowmobiles, skate wheels, motorcycle crank cases, bearings and electrical connections. The radiator in the 1982 model Ford Escort is molded from nylon-66.

The structure of dydactic nylons, such as nylon-ab is

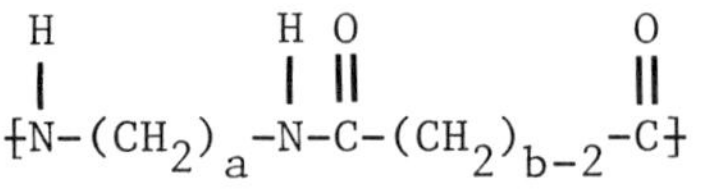

where a and b are equal to the number of carbon atoms in the repeating units of the diamine and dicarboxylic acid. Mono and biaxially oriented nylon film is available. (11)

Nylon-610 and nylon-612 produced by the condensation of hexamethylenediamine and sebacic or dodecanoic acid, respectively, are more resistant to moisture and more ductile than nylon-66. The properties of these polyamides may be improved by the formation of polyether blocks (NBC) and by blending with thermoplastics, such as EPDM, PET, PBT and TPE.

NBC (Nyrim) is more expensive than RIM polyurethane but it may be heated to 200°C without softening. NBC moldings are produced by the reaction injection molding (RIM) of polypropylene glycol and caprolactam in the presence of a catalyst. The tendency for this copolymer to swell in the presence of water is reduced by reinforcing with glass fibers.

Nylon-6 which has a heat deflection temperature of 80°C was produced by the ring opening polymerization of caprolactam in Germany in the 1940's. (12) Molded articles of these polymers may be produced in situ in the RIM process. (13) Nylon-11 and nylon-12 produced by the anionic polymerization of 11 and 12-amino acids are also characterized by good resistance to moisture and superior ductility. The structure of the repeating unit in monadic nylons, such as nylon-6 is:

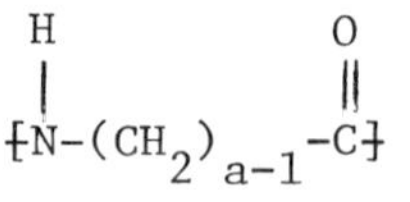

where a is the number of carbon atoms in the repeating unit. The copolymer of nylon-6 and nylon-66 is tougher and has a smoother surface than either of the homopolymers.

Injection moldable, yellow, transparent polyamide (PA7030) is also available. (14)

The aromatic polyamides (aramids) have been produced by the interfacial condensation of aromatic diamines, such as 1, 3-phenylenediamine and isophthaloyl chloride in chloroform. Amorphous transparent aramids with heat deflection temperatures of 160°C have been produced from 2,2-bis-4-(aminocyclohexyl) propane.(15)

Typical polyamides will have the following properties:

	Nylon-66	Nylon-6	PA7030	Aramid
tensile strength (psi)	12,000	18,000	11,000	17,500
elongation (%)	60	50	8	55
flexural strength (psi)	17,000	14,000	15,000	26,000
flexural modulus (psi)	420,000	390,000	-	640,000
izod impact (ft lbs/in of notch)	1.0	1.0	4.2	1.5
Rockwell hardness	R120	R119	-	M93
melting point (°C)	265	215	-	275
deflection temperature (°C)	75	65	105	250

Polycarbonates

The polycarbonates were discovered independently by Schnell(16) and Fox(17) in the 1950's. These tough engineering plastics are now produced at an annual rate of 130 thousand tons in the U.S. It is estimated that the annual production of these high performance plastics in the late 1980's will be 250 thousand tons. Polycarbonate and polycarbonate-polyester copolymers are being used for glazing, sealed beam headlights, door seals, popcorn cookers, solar heat collectors and appliances. The structure of the repeating unit in PC is

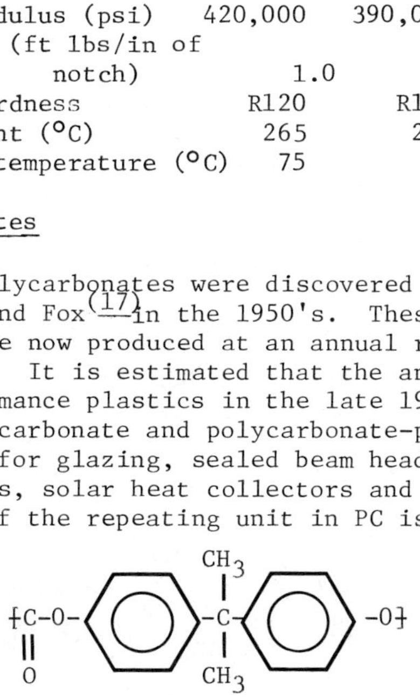

Polycarbonate (PC) tends to stress crack in the presence of gasoline but a 50-50 blend with polybutyl terephthalate (PBT, Xenoy) is unusually resistant to gasoline.(18, 19)

The properties of typical polycarbonates are as follows:

	unfilled	30% glass filled
tensile strength (psi)	9,500	19,000
elongation (%)	110	4
flexural strength (psi)	13,500	23,000
flexural modulus (psi)	340,000	1,250,000
izod impact (ft lbs/in of notch)	16	2
Rockwell hardness	M70	M92
melting point (°C)	150	tg=150
deflection temperature (°C)	120	140

Polyfluorocarbons

Polytetrafluoroethylene (PTFE Teflon) was discovered accidently by Plunkett(20) and commercialized by DuPont in the 1940's. This polymer has a solubility parameter of about 6H and a high melting point of 327°C and is not readily moldable. Polychlorotrifluoroethylene (CTFE, Kel-F), the copolymer of tetrafluoroethylene and hexafluoropropylene (FEP), polyvinylidene fluoride (PVDF, Kynar), the copolymer of tetrafluoroethylene and ethylene (ETFE), the copolymer of vinylidene fluoride and hexafluoroisobutylene (CM-1), perfluoroalkoxyethylene (PFA) and polyvinyl fluoride (PVF, Tedlar) are all more readily processed than PTFE. However, the lubricity and chemical resistance of these fluoropolymers is less than that of PTFE.

The structure of the repeating unit in PTFE is

$$+\overset{F_2}{\underset{}{C}} - \overset{F_2}{\underset{}{C}}+$$

The properties of typical fluoropolymers are shown below:

	PTFE	FEP	PVDF
tensile strength (psi)	3,000	2,800	6,000
elongation (%)	200	250	100
flexural strength (psi)	-	-	900
flexural modulus (psi)	60,000	-	20,000
izod impact (ft lbs/in of notch)	13	-	4
Rockwell hardness	-	D60	-
melting point (°C)	327	275	156
deflection temperature (°C) (66psi)	120	-	80

Polyphenylene oxide

Injection moldable polyphenylene oxide (PPO, Noryl) was produced by A. S. Hay in the early 1960's by the copper-amine catalyzed oxidation of xylenols.(21) The commercial product is a blend of PPO and polystyrene. PPO is being produced at an annual rate of 70 thousand tons in the U.S. and it is anticipated that the annual production in the late 1980's will exceed 175 thousand tons.

The structure of the repeating unit in PPO is

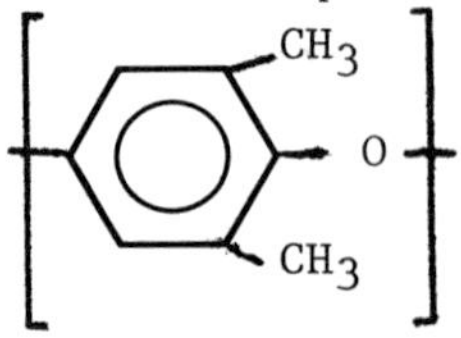

PPO extruded sheet is being used for solar energy collectors, life guards on broadcasting towers, airline beverage cases and window frames. The properties of a typical PPO-PS blend are as follows:

	PPO-PS	30% glass filled
tensile strength (psi)	9,500	18,000
elongation (%)	60	4
flexural strength (psi)	13,000	21,000
flexural modulus (psi)	355,000	20,000,000
izod impact (ft lbs/in of notch)	5	2
Rockwell hardness	R115	R115
glass transition temperature (Tg)	135	105
deflection temperature (°C)	120	130

Polyphenylene sulfide

Polyphenylene sulfide (Ryton) is produced by a Wurtz Fittig reaction of p-dichlorobenzene and sodium sulfide. (22) This high melting (290°C) crystalline polymer may be injection molded or processed as a powder by compacting, sintering and forging. The compressability of the polymeric powder is improved by the addition of PTFE filler. PPS moldings may be crosslinked by heating at temperatures above 370°C.

PPS is used at an annual rate of about 2,000 tons but is anticipated that this rate will increase to 25 thousand tons in the late 1980's. This polymer is being used for pumps, sleeve bearings, cookware, quartz halogen lamp parts and electrical appliances. (23)

The structure of the repeating unit in PPS is

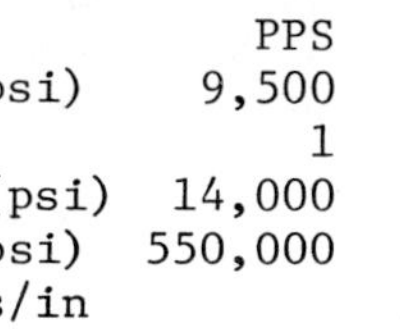

The properties of typical PPS molded products are as follows:

	PPS	PPS with 40% glass
tensile strength (psi)	9,500	19,500
elongation (%)	1	1
flexural strength (psi)	14,000	29,000
flexural modulus (psi)	550,000	1,700,000
izod impact (ft lbs/in of notch)	-	-
Rockwell hardness	R123	R123
melting point (°C)	290	290
deflection temperature (°C)	135	250

Polyaryl esters

Thermoset polyesters, such as Glyptal, alkyds and glass reinforced unsaturated polyesters have been available for several decades but injection moldable polyaryl esters are relatively new. Whinfield and Dickerson extruded polyethylene terephthalate (PET) fibers (Dacron) in the 1940's.[24] PET was also extruded as film (Mylar) in the 1950's but this polymer was not blow molded commercially until the 1970's.[25]

The addition of antinucleating agents permits the injection molding of PET (Rynite). Injection moldable glycol-modified polyesters (PETG, Kodar) have been produced by partial replacement of the ethylene glycol by cyclohexanol dimethylol.

The structure of the repeating unit in PET is

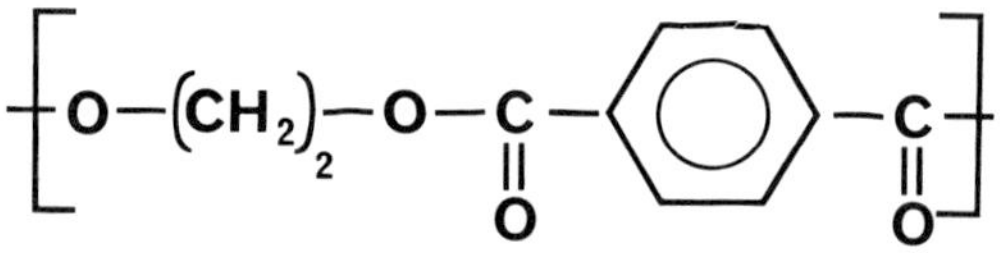

Moldability of aryl polyesters have also been improved by the use of polybutylene terephthalate (PBT) instead of PET or by the use of blends of PET and PBT. PBT under the trade name of Celanex, Valox, Gafite and Versel is being produced at an annual rate of 25 thousand tons. Copolymers of carbonate and aryl esters, acrylics and aryl esters, and imide and aryl esters as well as physical blends of polyesters and other polymers are available. These aryl polyesters are being used for bicycle wheels, springs, and blow molded containers. The properties of typical aryl polyesters are as follows:

	PET	PBT	PBT 30% glass filled
tensile strength (psi)	8,500	8,500	12,000
elongation (%)	50	50	4
flexural strength (psi)	14,000	12,000	26,000
flexural modulus (psi)	350,000	300,000	1,100,000
izod impact (ft lbs/in of notch)	0.3	0.8	1.5
Rockwell hardness	M94	M70	M90
melting point (^{o}C)	245	235	245
deflection temperature (^{o}C)	50	40	105

A new high impact blend of PBT and polybutene (Valox CT) is also available.[26] The melt viscosity of blends of PET and nylon-66 has been reduced by the addition of polyvinyl alcohol. Self reinforcing PET has been produced by the addition of p-hydroxybenzoic acid which forms liquid crystals in the composite.

Polyaryl sulfones

There are several types of polyaryl sulfones available commercially. The classic polysulfone (Udel) was produced by the nucleophilic displacement of the chlorine substituents on bis(p-chlorophenyl) sulfone by the sodium salt of bisphenol A.(27) Another polysulfone (Astrel) is produced by the Friedel-Crafts condensation of biphenyl with oxy-bis(benzene sulfonyl chloride). Another polysulfone (Vitrex), is produced by the alkaline condensation of bis(chlorophenyl) sulfone.(28) Blends of polysulfones with ABS (Arylon, Mindel) and SAN (Ucardel) are available.

Polysulfones which are produced at an annual rate of 20 thousand tons are being used for ignition components, hair driers, cookware and structural foams.(29)

The structure of the repeating unit in a typical polyaryl sulfone is

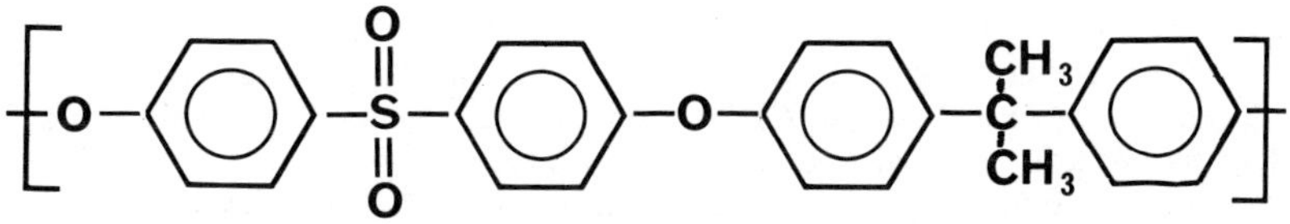

The properties of typical polyaryl sulfones are as follows:

	Polyaryl Sulfone	Polyaryl Sulfone +30% Fiber glass
tensile strength (psi)	6,500	15,000
elongation (%)	50	2
flexural strength (psi)	15,000	30,000
flexural modulus (psi)	360,000	1,350,000
izod impact (ft lbs/in of notch)	1.2	1.1
Rockwell Hardness	M69	M90
glass transition temp. (°C)	190	-
deflection temperature (°C)	175	185

Gruman has developed an automated triangular truss-type beam builder using graphite-reinforced polyether sulfone. This beam can be formed in outerspace at the rate of 1.5m per min. from flat stock which is heated and forced continuously around a die. The sections are induction welded and may be protected from deterioration by the application of a coating of polyether ether ketone (PEEK).(30)

Polyether ether ketone

ICI has introduced a new crystalline polyether ether ketone

(PEEK), which has a glass transition temperature of 145°C and a heat deflection temperature for reinforced PEEK of 300°C.[31] PEEK is being considered for use as blow molded containers for nuclear wastes, jet engine components, printed circuits, electrical applications and wire coatings.[32]

The structure of the repeating unit in PEEK is

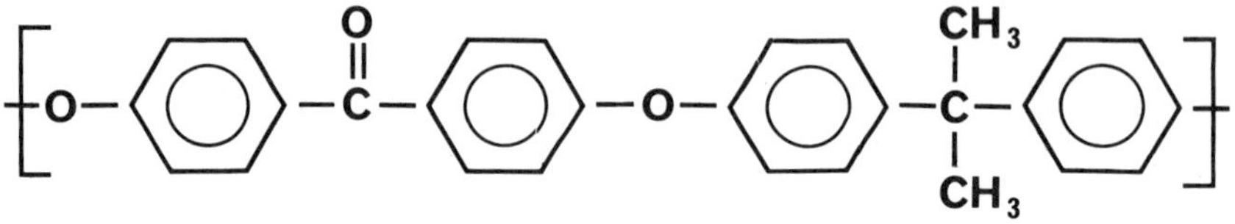

Polyimides

Heat resistant intractable thermoset polyimides (Vespel, Kinel, Kapton) have been supplemented by injection moldable polyamide-imides (Torlon) and modified polyimides (Kanox, Toramid).[33] Thermoplastic polyimide adhesives were cited as one of the top 100 inventions in 1981. Polyimide foam does not ignite at temperatures below 430°C and is being considered as a cushioning material in public conveyances.

The structure of the repeating unit in polyimide is

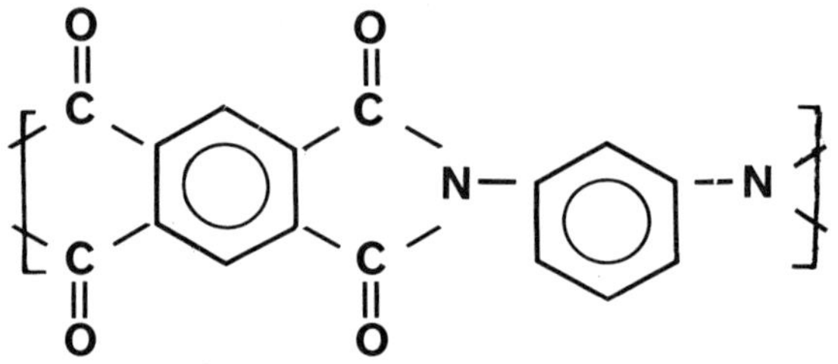

The processing properties are improved when polyimides are produced from meta substituted diamines.[34] A new amorphous polyether-imide (Ultem) with a high heat deflection temperature has been introduced by the General Electric Co.

The ether linkages in PEI improve the ease of processing and increase the ductility of these high performance plastics.[35] PEI reinforced with 30% fiberglass has a heat deflection temperature of 210°C.[36]

The polyamide-imide (Torlon, PA-1) has a heat deflection temperature of 285°C. A Ford automobile prototype engine has been built from PA-1.[37, 38]

Conclusions

In spite of their utility and versatility, the new high performance polymers will not displace general purpose thermoplastics, classic thermosets, ceramics or metals unless there is a demonstratable superiority. Fortunately, there are many applications for high performance thermoplastics for which no other material is adequate. Hence, the annual production of high performance thermoplastics in the late 1980's will exceed the production of all plastics in the 1940's. This unprecedented growth of these engineering polymers will be based entirely on their performance.

Literature Cited

1. Seymour, R.B., Ed. "Plastic Mortars, Sealants, and Caulking Compounds"; No. 113 ACS Symposium Series, American Chemical Society: Washington, DC, 1979.
2. Seymour, R.B., Ed. "Additives for Plastics"; Vols I and II, Academic Press: New York, 1978.
3. Seymour, R.B.; Kispersky, J.P.; U.S. 2,439,227
4. Seymour, R.B., Ed. "History of Polymer Chemistry"; Marcel Dekker: New York, 1982.
5. McDonald, R.N.; U.S. 2,768,994
6. Walling, C.; Brown, F.; Bartz, K.W.; U.S. 3,027,352
7. Seymour, R. B.; "Plastics vs Corrosives"; John Wiley and Sons: New York, 1982.
8. Carothers, W.H.; U.S. 2,130,947
9. Garshman, A. Plast Technol, 1978, 24(11), 79.
10. Swanson, D.E. Plast Des and Proc, 1978, 17(5), 44.
11. Wood, A.S. Modern Plastics, 1982, 58(8), 46.
12. Reinschuessel, H. J. J. Poly Sci/Macromolecules, 1977, 12, 65.
13. Riley, M. W. Plast Technol, 1979, 25(4), 11.
14. Riley, M. W. Plast Technol, 1982, 28(6), 15.
15. Hill, R. Polym Eng Sci, 1978, 18, 36.
16. Schnell, H. Angew Chemie, 1956, 68(2), 633.
17. Fox, D. W.; Chap 310 "Applied Polymer Science"; Craver, J.K. and Tess, R.W. Eds. American Chemical Society: Washington, DC, 1975.
18. Riley, M. W. Plast Technol, 1982, 28(6), 29.
19. Gross, S. Modern Plastics, 1982, 59(6), 34.
20. Plunkett, R.J.; U.S. 630,654.
21. Fox, D. W.; Chap 30 "Applied Polymer Science"; Craver, J.K. and Tess, R. W. Eds. American Chemical Society: Washington, DC, 1975.
22. Bolke, P. U. U.S. Dix Plast Eng, 1982, 38(2), 25.
23. Edmunds, J.T.; Hill, H. W.; U.S. 3,334,129.
24. Whinfield, J. R. Nature, 1945, 138, 930.
25. Pengilly, W. O. Modern Plastics, 1978, 55(104), 50.

26. Gross, S. Modern Plastics, 1982, 59(2), 14, 74.
27. Leslie, V. J.; Rose, J.; Rudken, G.O.; Feltzen, J. Chem Tech, 1975, 5, 426.
28. Von Hassel, A. Plast Technol, 1980, 25, 57.
29. Steming, J.; Smith, C.P.; Kimber, P.J. Modern Plastics, 1981, 58(11), 86.
30. Riley, M. W. Plast Technol, 1981, 27(8), 139.
31. Riley, M. W. Plast Technol, 1981, 28(6), 21.
32. Riley, M. W. Plast Technol, 1981, 27(8), 139.
33. Bitterbeck, C. J. Modern Plastics, 1981, 57(10A), 46.
34. St. Clair, A.K.; St. Clair, T.L. SAMPE Quart, 1981, 13(3), 20.
35. Rimsa, S.B.; Serfaty, I.W., presented at SPE-Antec, San Francisco, CA, May 10, 1982.
36. Riley, M. W. Plast Technol, 1982, 28(3), 17.
37. Floryan, D. E.; Serfaty, I. W. Modern Plastics, 1981, 59(6), 146.
38. Spaulding, L.D. Plast Eng, 1981, 28(3), 25.

RECEIVED June 13, 1983

8

Model Fungicidal Monomers, Polymers, and Latices for Paints

CHARLES U. PITTMAN, JR.

Chemical Consultant, 132 Woodland Forrest, Tuscaloosa, AL 35405

G. ALLAN STAHL

Phillips Petroleum Company, Research and Development, Bartlesville, OK 74004

Summary

When a biocide is simply mixed into a paint it may subsequently be lost from the painted surface by leaching or vaporization. To overcome this problem several biocidal compounds have been converted to vinyl monomers and subsequently incorporated into homo-, co-, and terpolymers. Latices have been prepared for several systems. Since mildew defacement of painted surfaces is widespread problem, we have concentrated on polymer anchoring of fungicidal compounds and compounds with wide biocidal activity. These included pentachlorophenol, 8-hydroxyquinoline, 2-(4'-thiozoyl)benzimidazole, 3,4',5-tribromosalicylanilide, o-benzyl-p-chlorophenol, o-acrylyl salicylanilide and 2-mercapto-pyridine-N-oxide. The reactivity ratios of the biocidal monomer, pentachlorophenyl acrylate, were obtained with both vinyl acetate and ethyl acrylate. While it is possible that some biocides may function when incorporated into a macromolecule, it seems likely that many biocides must be released from the polymer to be active. Thus, the biocidal monomers were incorporated into the polymers using functions with different hydrolytic propensities. These included, fungicidal acrylates, 2-fungicidalethyl acrylates, and fungicidal vinyl ethers. The polymerization behavior of several such monomers was examined. Stable terpolymer latices were prepared from methyl methacrylate, n-butyl acrylate and each of the following fungicidal monomers: (1) pentachlorophenyl acrylate, (2) 2-pentachlorophenoxyethyl acrylate, (3) 2-(8-quinolinyloxy)ethyl acrylate, (4) acryloyloxy-3,4',5-tribromosalicylanilide, (5) 2-(2-acryloylethoxy)-3,4',5-tribromosalicylanilide, (6) 2-(o-benzyl-p-chlorophenoxy)ethyl acrylate and (7) vinyl o-benzyl-p-chlorophenyl ether. A terpolymer latex of vinyl acetate, 2-ethylhexyl acrylate and acryloyloxy-3,4',5-tribromosalicylanilide was made. A latex containing pentachlorophenol bound by two different releasing groups and a latex with two different fungicidal monomers was made.

0097-6156/83/0229-0099$10.75/0
© 1983 American Chemical Society

Accelerated growth testing of monomers, polymers, and latices containing chemically bound fungicides demonstrated that the concept of polymer-anchoring biocides has promise. Several systems were shown to control microorganism growth in minimum inhibitory concentration tests using both agar dilution and aqueous nutrient broath methods. A variety of in vitro agar dish growth tests and the "Zabel Test" were also employed. Photochemical cleavage of biocide from some polymers was demonstrated in UV irradiation agar dish tests. Taken together the results provide laboratory support for the concept that binder polymers, containing chemically bound biocides, can be useful for controlling microorganism growth.

Introduction

Mercury-containing mildewcides have come into disfavor due to the high toxicity of methyl mercury compounds to humans. Since mercury salts have been widely used for bacterial and fungal control in paints more acceptable replacements have been increasingly sought in recent years. A research program sponsored by the Paint Research Institute at the University of Alabama was undertaken to look for organic biocides which could replace mercury salts. Since organic compounds may leach or vaporize from thin, high surface area paint films, the focus of the Alabama program became the development of polymer-anchored fungicides. A detailed literature search[1] in 1976 revealed that little work had been done with this concept in connection with outdoor coatings except for some studies of trialkyltin esters[2]. Drisco and coworkers[2-5] prepared paint resins containing tri-n-butyltin acrylate. Laboratory tests indicated these resins were resistant to Aureobasidium pullulans, Aspergillus niger, and Aspergillus oryzae, and that fungicide leaching from the coating was considerably reduced compared to the use of the blended fungicide. However, field tests of these coating resins[4,5] on pine and redwood test panels began to show microorganism growth after three months.

Mildew defacement of organic coatings has long been a major problem of the paint industry. Aureobasidium pullulans is the major causitive organism responsible for this defacement.[6,7] A current approach to solve this problem is to blend a fungicide into the paint formulation. Although this solution has had success for short time periods, the fungicide tends to be vaporized or leached from the coating over long time periods. After the concentration of fungicide drops below critical levels, mildew may start growing on the coatings surface.[8]

Several factors must be considered in selecting a fungicide for blending into paints.[9] The four most important factors, other than cost, are toxicity, solubility, volatility, and UV stability. Toxicity to fungi and bacteria must be high, but toxicity to humans should be low. The solubility of any candidate fungicide must be carefully considered for several reasons. The fungicide must not be water soluble since it would leach from coatings during rain. Even a slight water solubility eliminates many fungicides for selection in paints. Also, the fungicide needs to be soluble in the paint or to be dispersed without settling or agglomerating. If the fungicide agglomerates, dispersing agent use in the paint may be required leading to formulation complications and higher costs. Fungicides with appreciable vapor pressures cannot be considered for paint use since they would vaporize from the paint after application to a surface. Ultraviolet photodecomposition may degrade certain fungicides and such fungicides must be avoided.

The concept of polymer-anchoring a biocide promises several advantages. Volitile biocides may be considered because they would be chemically attached to the paint binder. Water-soluble biocides also may be considered since the polymer they would be anchored to would be water insoluble. The polymer-anchored biocide would be molecularly dispersed throughout the film rather than present in small particles. Since the biocide would remain within the film longer if it were polymer anchored, a lower level of biocide incorporation in the paint might be possible for a given useful mildewcidal lifetime. Also, the anchored biocides would be less toxic to humans which may permit the use of certain compounds which otherwise would be avoided. However, with these advantages come several potential problems. The polymerization of potential biocidal monomers could be troublesome. The incorporation of such monomers into the polymer would have to be kept at low mole percents to avoid changing substantially the polymer's properties (i.e., Tg, etc.). Proper design of copolymers or terpolymers could overcome this problem.

A fundamental question involves the biocidal activity of the biocide when it is chemically attached to a polymer. Will the polymer be an active biocide? If the biocidal activity results from an interaction of the biocide at the cell wall or by inactivating an exocellular enzyme, then it is possible that the polymer, itself, could be an active biocide. However, if the biocide must be incorporated into the organism to function, then it will have to be cleaved from the polymer binder prior to exhibiting activity. For that reason polymers having biocides attached by functional groups of differing hydrolytic susceptability were made.

Design of polymer-bound fungicides can accommodate a slow release mechanism or a permanently attached biocide. The chemical linkages used to attach fungicides to polymers could either be hydrolyzable or nonhydrolyzable depending on the type chemical bonding used. Ester or amide linkages, which undergo acid-, base-, or enzyme-catalyzed hydrolysis, may be used to bind the fungicide and, therefore, later release the free fungicide upon cleavage of that linkage. These linkages could be designed to cleave at various rates under environmental conditions (or by the enzymes released by microorganisms). If environmental conditions cause hydrolysis of the linkage, there would be some loss of fungicide due to leaching, but the rate of leaching could be far less than the leaching rate of that same fungicide if it were simply blended into the paint. Many fungicides might function to retard growth only after they have been released from the coating, but this may not be true of all fungicides. Once released from the polymer, the concentration of free fungicides might be too low to achieve the desired result. Nonhydrolyzable linkages, such as ether or hydrocarbon bonds, could also be used. These linkages would stop mildew growth only if the fungicide is still active when it is a part of the polymer. To our knowledge, no studies of fungicides bound to polymers by these types of linkages have been done. Slow release herbicides which use the polymer-anchoring concept have been developed.[10,11]

Over the past few years several biocides have been converted to monomers containing ester, amide, or ether functions which bind the biocide and the polymerization of these monomers was studied.[12-16] The biocides examined included pentachlorophenol, 1,[12-16] 8-hydroxyquinoline,2,[12,14-16] 3,4',5-tribromosalicylanilide,3,[12, 14-16] o-benzyl-p-chlorophenol,4,[14-16] salicylanilide,5,[12] 2-(4'-thiazoyl)benzimidazole,6,[12] and 2-mercaptopyridine-N-oxide,7.

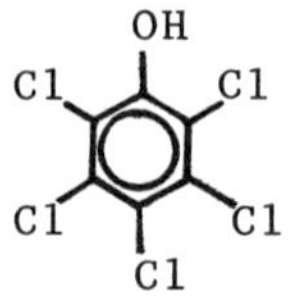

Pentachlorophenol

1

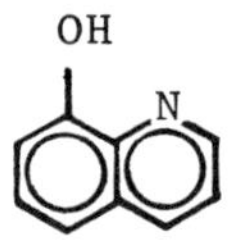

8-Hydroxyquinoline

2

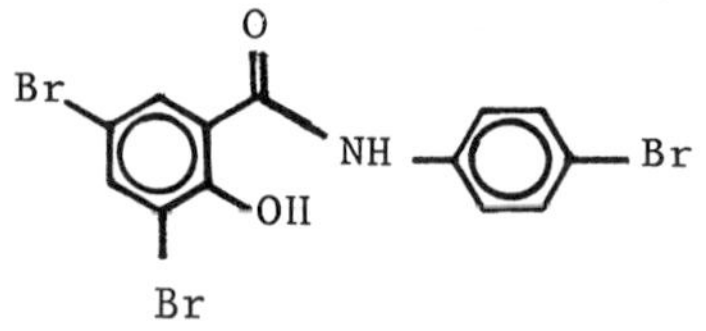

3,4',5-Tribromosalicylanilide

3

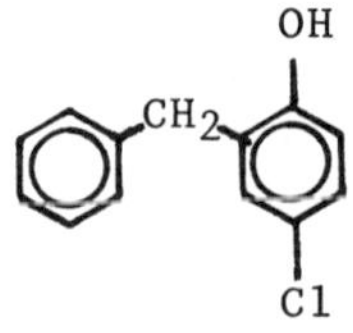

o-Benzyl-p-chlorophenol

4

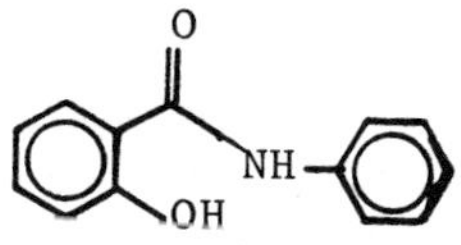

Salicylanilide

5

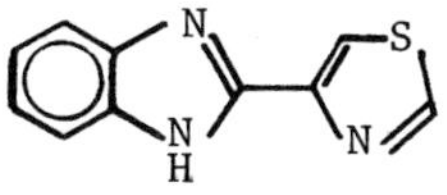

2-(4'-thiazoyl)benzimidazole

6

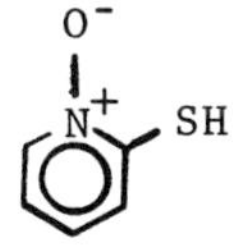

2-Mercaptopyridine-N-oxide

7

Pentachlorophenol was chosen because it is a broad spectrum biocide widely used as a wood preservative and as a slime control agent.[17-20] 8-Hydroxyquinoline had been used as a fungicide before 1900 and is thought to function by chelating metal ions.[21-23] Tribromosalicylanilide is an active biocide with low dermatological effects to humans.[24-26] Its biocidal effects vary upon the addition of surfactants to its medium.[25, 26] o-Benzyl-p-chlorophenol is known to be a broad spectrum commercial bacteriacide[27,28] but its fungicidal properties have not been published. 2-(4'-thiazoyl)benzimidazole is a fungicide, produced by Merck Inc., which has been used in paints. Due to its nonvolitility and low solubility 6 can be mixed with paints and remain in the film. 2-Mercaptopyridine-N-oxide is a fungicide produced by Olin Inc. and used widely in shampoos as an antidandruff agent.

Results

Monomer Preparation

Each of the fungicides, 1-5 has a phenolic hydroxy group. Therefore, the acrylic esters of each are readily prepared. However, the stable phenoxide anions of 1-5 are good leaving groups. Thus, these acrylic esters or their polymers should undergo reasonably facile hydrolysis. For this reason, polymers of these acrylates should be able to release the specific biocide as shown in Scheme I. The hydrolysis of 8-hydroxyquinoline esters is particularly rapid due to neighboring group participation of nitrogen.[12,1]

To provide a series of monomers (polymers) which would hydrolyze more slowly than the above-mentioned phenolic acrylates, the phenolic group was first converted to its 2-hydroxyethyl ether derivative which, in turn, was converted to its corresponding chain-extended acrylate (see Scheme II). Ester hydrolysis of these chain-extended esters releases an alkoxide anion which is a stronger base, hence a poorer leaving group, than phenoxy anions. Thus, polyacrylates of these monomers should release the biocidal moiety more slowly than the phenolic polyacrylates. Further, it should be noted that the 2-hydroxyethyl ether of the original fungicide is released. This may be biologically inactive or less active than its parent fungicide (i.e. 1-5). These ethers may (or may not) have to further degrade back to the parent fungicide to display fungicidal activity. Binder polymers containing both the acrylate and chain extended acrylates might exhibit extended protection against mildew defacement.

A further task was the construction of polymers from which a bound fungicide would not hydrolyze. Since aryl secondary alkyl ethers are only cleaved under vigorous conditions, we selected the vinyl ethers of fungicides 1-4 as target monomers. A disadvantage recognized at the outset was the expected difficulty of copolymerizing vinyl ethers (which undergo ready cationic polymerization) with acrylic monomers which are polymerized using radical or anionic initiators but which resist cationic initiation (Scheme III).

All of the acrylate and chain-extended acrylate syntheses[12, 14-16] are summarized in Scheme IV together with yields. Scheme IV also illustrates the route that was used[12] to convert _6_ to its acylamide analog 2-(4'-thiazoyl)benzimidazoyl acrylate,_21_. This monomer polymerizes to give polymers[12] which undergo facile hydrolysis to free _6_ because the 2-(4'-thiazoyl)benzimidazolyl leaving group is a stable anion (by virtue of resonance stabilization of negative charge by the benzimidazole ring). This ready hydrolysis was illustrated by the inability to prepare the monomer on treatment of _6_ with acryloyl chloride[12]. Thus, the sodium salt of _6_ was made and reacted with acryloyl chloride but even this method resulted in losses on work up and a poor overall yield (16-17%) of _21_.[12]

SCHEME I

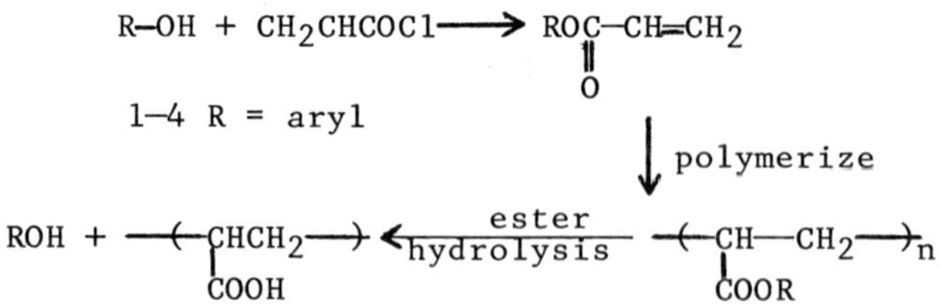

SCHEME II

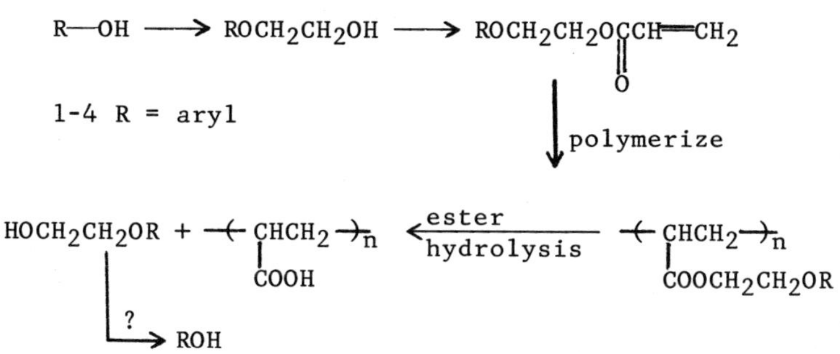

SCHEME III

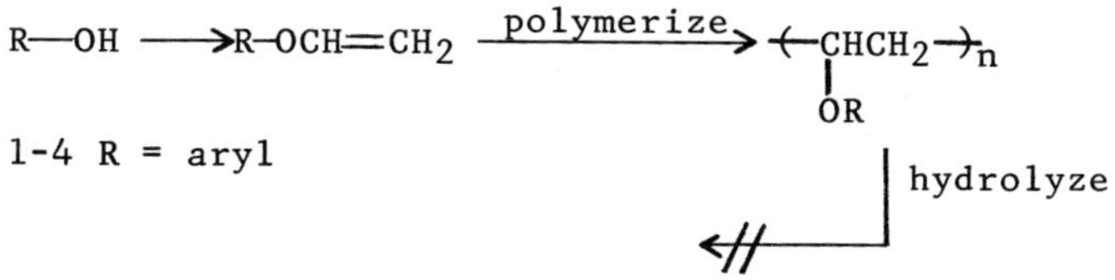

SCHEME IV

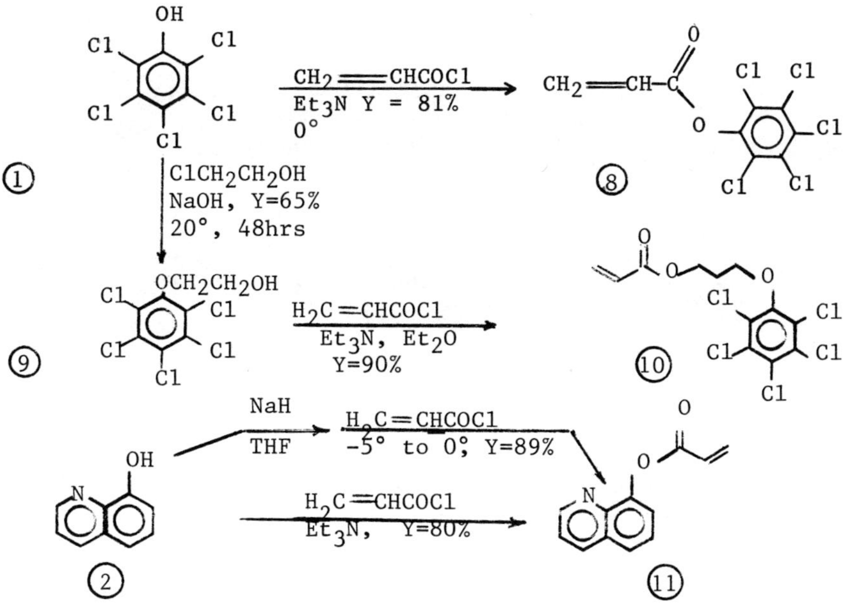

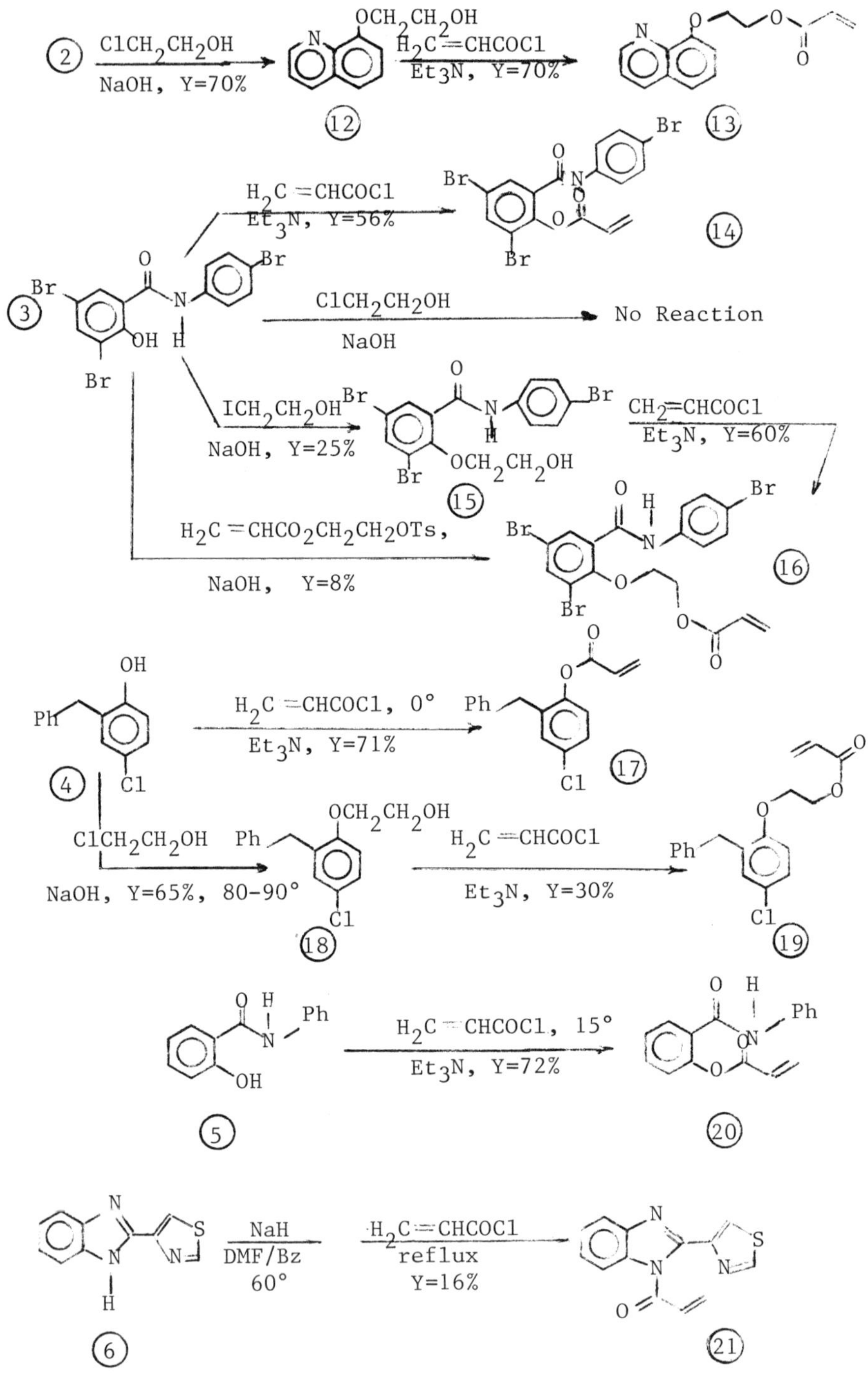

2
ClCH2CH2OH
NaOH, Y=70%
OCH2CH2OH
12
H2C=CHCOCl
Et3N, Y=70%
13
Br
3
H2C=CHCOCl
Et3N, Y=56%
14
ClCH2CH2OH
NaOH
No Reaction
ICH2CH2OH
NaOH, Y=25%
15
CH2=CHCOCl
Et3N, Y=60%
H2C=CHCO2CH2CH2OTs,
NaOH, Y=8%
16
4
Ph
Cl
H2C=CHCOCl, 0°
Et3N, Y=71%
17
ClCH2CH2OH
NaOH, Y=65%, 80-90°
18
H2C=CHCOCl
Et3N, Y=30%
19
5
H2C=CHCOCl, 15°
Et3N, Y=72%
20
6
NaH
DMF/Bz
60°
H2C=CHCOCl
reflux
Y=16%
21

Several attempts to make the thiol acrylate of 7 by treating 7 with acryoylchloride or acryoyl anhydride failed to give the desired product. This material is probably unstable due to the nucleophilic effect of the nearby oxygen on nitrogen. Presumably a mixture of readily hydrolyzed O- and N-substituted derivatives was formed. Thus, an attempt was made to form the thiohemiacetal with formaldehyde. Unexpectedly, two moles of formaldehyde condensed to give hemiacetal, 22 which was then converted to acrylate 23 (Scheme V).

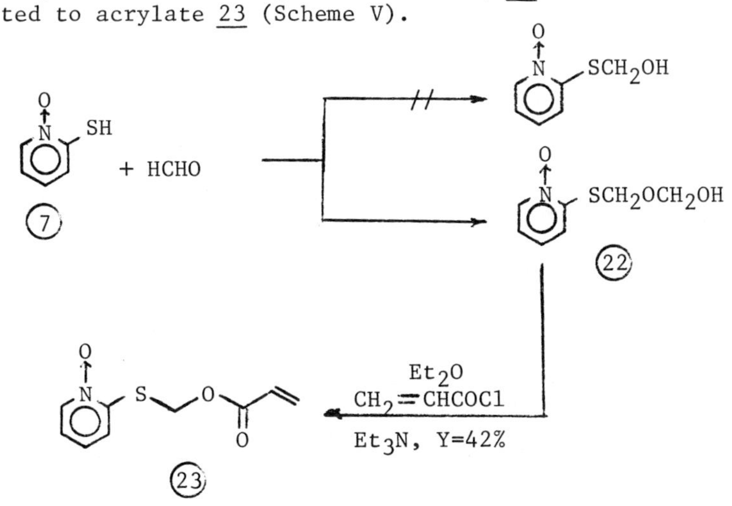

SCHEME V

Acrylate and Extended-Chain Acrylate Synthesis[12,14-16]

The syntheses of both acrylate and extended-chain acrylates of fungicides 1-5 are summarized together with yields in Scheme IV. Acrylates 8, 11, 14, 17, and 20 were prepared in good yields using acryloyl chloride under normal Shotten-Baumann conditions.[12,14] This method worked well even for the sterically hindered 3,4',5-tribromosalicylanilide. Syntheses of chain-extended acrylates 10, 13 and 19 were accomplished by reacting fungicides 1, 2, and 4, respectively, with 2-chloroethanol and NaOH to first generate the 2-fungicidalphenoxyethanols[14,16] (i.e. 9, 12 and 18). Treatment of these chain-extended alcohols with acryloyl chloride under Shotten-Baumann conditions gave the chain-extended acrylates 10, 13, and 19. This route was unsuccessful, however, starting with 3,4',5-tribromosalicylanilide because displacement of chloride from 2-chloroethanol failed to give the chain-extended alcohol.[14] The phenoxy oxygen in 3,4',5-tribromosalicylanilide is severely hindered by the ortho bromo and amide functions, and this hindrance retards the rate of S_N-2 displacement. By using 2-iodoethanol, a more reactive substrate to S_N-2 displacements, the chain-extended alcohol 15 was obtained albeit in low (25%) yield.[14] Esterification of 15 with acryloyl chloride gave the chain-extended acrylate 16.[14] The structures of each of the monomers was confirmed by infrared and nuclear magnetic resonance spectroscopy.

Vinyl Ether Syntheses

Three general synthetic routes were examined for the preparation of the vinyl ether monomers of fungicides 1, 3, and 4.[14] These results are summarized in Scheme VI. First, the 2-bromoethyl ethers 24, 27, and 29 were prepared from 1, 3, and 4, respectively, via nucleophilic displacement of bromide from 1,2-dibromoethane in the presence of NaOH or KOH. Then, base-catalyzed dehydrohalogenation gave the desired vinyl ether monomers 25, 28 and 30 in satisfactory yields.[14] A second route, based on literature precedent,[29,30] involved treating the fungicidal phenols 1, 3, and 4 with vinyl acetate in the presence of mercuric acetate to give the vinyl ether in an addition-elimination sequence.[14] This method did not work on sterically hindered 3 and it gave the acetal 31 (and not vinyl ether 30) when employed with biocide 4. The third route was based on the use of 1,2-vinylbis(triphenylphosphonium)dibromide.[31] Reaction of the fungicides 1 and 4 with this salt in chloroform and triethylamine, followed by treatment with aqueous NaOH, gave vinyl ethers 25 and 30.[14] This method was not successful in generating the vinyl ether 28 from tribromosalicylanilide.[14] 8-Hydroxyquinoline was readily converted to vinyl ether 32 in good yields[14] by treatment with acetylene in base at elevated temperatures and pressures using methods similar to previous reports.[32-34]

Known Biological Activity of the Monomers

It is important to consider the biocidal properties of the monomers as well as the parent fungicides since unreacted monomer may remain in a latex after its preparation. Pentachlorophenyl acrylate, 8, and its homopolymer both inhibit the growth of *S. aureus*, *A. niger*, *Penicillium citrinum*, and *Saccharomyces cervisiae*[35] The polymer was less active. Poly(vinyl chloride) compositions are protected against microbial attack using 8 in concentrations of 1-2% by weight.[36] 2-Pentachlorophenoxyethyl bromide, 24, is known to be fungistatic[37] and 2-pentachlorophenoxyethanol, 9, has a fairly high fungicidal activity but it is less toxic (to mice) than pentachlorophenol.[37] Extended-chain acrylate 10 has not been previously reported. 8-Quinolinyl acrylate, 11, has only been tested in the control of *Alternaria solani* on tomatoes.[38,39] Vinyl-8-quinolinyl ether, 32, in the form of iron, tin, gold, and zinc complexes were active against Staphylococcus[33] but not as active as 8-hydroxyquinoline, 2. 2-(8-Quinolinyl)ethanol, 12 is also less active than 2 but was fairly effective against *Staphyloccus pyrogenes* and *S. aureus*.[40]

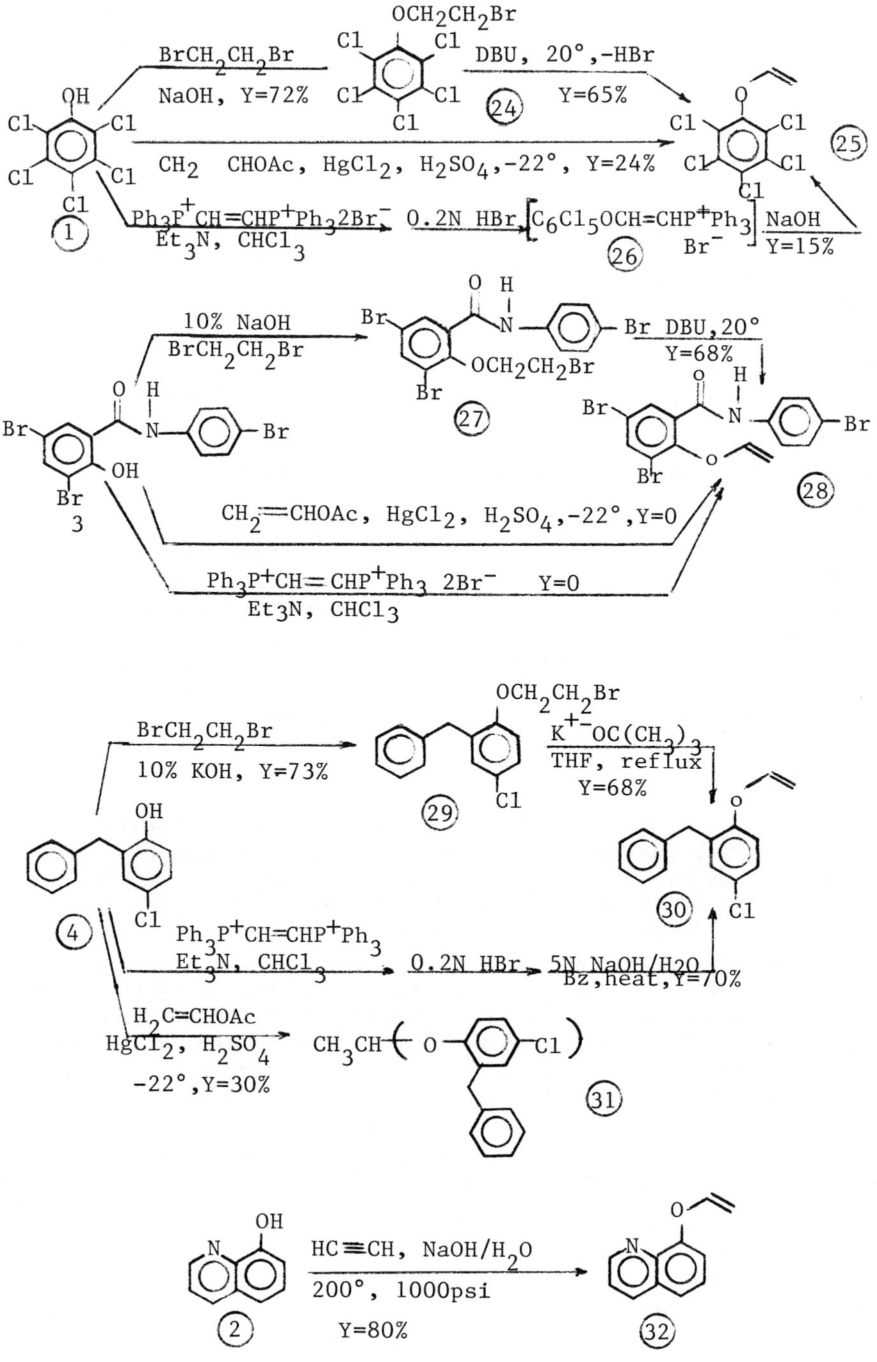

OCH2CH2Br
BrCH2CH2Br
NaOH, Y=72%
DBU, 20°,-HBr
Y=65%
24
25
1
CH2 CHOAc, HgCl2, H2SO4,-22°, Y=24%
Ph3P+CH=CHP+Ph32Br-
Et3N, CHCl3
0.2N HBr
[C6Cl5OCH=CHP+Ph3] Br-
26
NaOH
Y=15%
10% NaOH
BrCH2CH2Br
OCH2CH2Br
27
DBU, 20°
Y=68%
28
3
CH2=CHOAc, HgCl2, H2SO4,-22°,Y=0
Ph3P+CH=CHP+Ph3 2Br- Y=0
Et3N, CHCl3
BrCH2CH2Br
10% KOH, Y=73%
OCH2CH2Br
29
K+-OC(CH3)3
THF, reflux
Y=68%
30
4
Ph3P+CH=CHP+Ph3
Et3N, CHCl3
0.2N HBr
5N NaOH/H2O
Bz,heat,Y=70%
H2C=CHOAc
HgCl2, H2SO4
-22°,Y=30%
31
HC≡CH, NaOH/H2O
200°, 1000psi
Y=80%
2
32

SCHEME VI

The extended chain acrylate 13 has not been reported. No reports exist concerning biological activity of monomers 14, 16, or 28, derived from tribromosalicylanilide, or 17, 19, or 30 derived from o-benzyl-p-chlorophenol.

Polymerizations[12,14,16]

Fungicidal acrylates 8, 11, 14, and 17 were homopolymerized in solution using radical initiation (AIBN).[14] Only very low yields and low molecular weights were obtained with 17, presumably due to the high chain transfer constant expected from the dibenzylic methylene function within this monomer.[14] Anionic initiation of 17 with $LiAlH_4$ in THF also failed. A 24% yield of a low molecular weight homopolymer of 17 resulted from thermal polymerization at 150°C. Good yields and high molecular weights could be achieved in the homopolymerizations of 8, 11, and 14. Chain-extended acrylates 10, 13, 16, and 19 resisted homopolymerization under AIBN initiation. Sample homopolymerization results are summarized in Table 1.[14]

Acrylates 8, 14, and 17 were also copolymerized with n-hexyl methacrylate or n-butyl acrylate in emulsion reactions (sodium lauryl sulfate, $K_2S_2O_8$, 60°C) as were chain-extended acrylates 10, 13, 16 and 19.[14] The copolymers of 17 and 19 were low molecular weight materials again due to easy chain transfer. The bulk copolymerization of 19 with n-hexyl methacrylate gave reasonably high molecular weights.[14] Some solution copolymerizations were also carried out. Representative results are summarized in Table 2.

A series of film forming vinyl acetate or ethyl acrylate copolymers were prepared where the fungicidal monomers was pentachlorophenyl acrylate, 8, 8-quinolyl acrylate, 11, or acrylyl salicylanilide, 20.[12] In these copolymers the mole fraction of the fungicidal monomer was kept low (i.e. between 0.04 and 0.11). Representative copolymers are summarized in Table 3.[12] These copolymers were prepared by bulk copolymerization using the free radical initiators AIBN (with 8) and t-butylperoxypivalate (BPP) (with 11 and 20). The copolymer compositions were determined both by elemental analyses and pyrolysis gas chromatography.[12] As expected with bulk polymerizations, broad molecular weight distributions were obtained. As can be seen in Table 3, the molecular weights were high. The glass transition temperatures increased as the mole content of the fungicidal monomer increased (Table 3).[12] The sensitivity of Tg to small incorporations of 8, 11, or 20 indicates the homopolymers of these monomers would have high Tg values.

Bulk copolymerizations of 2-(4'-thiazoyl)benzimidazole acrylate, 21, with ethyl acrylate were successful.[12] However, attempts to copolymerize 21 with vinyl acetate in bulk gave only

Table 1-Representative Homopolymerization Reactions Of Fungicidal Monomers

Fungicidal Monomer	Solvent	Initiator	Conditions °C/time, hr	Yield (%)
8	Benzene	AlBN	60/36	87[a]
14	Benzene	AlBN	70/24	75[b]
17	Benzene	AlBN	60/20	0
17	THF	$LiAlH_4$	-78→20/14	0
17	neat	------	150/8	24
10	Benzene	AlBN	65/51	0
13	Benzene	AlBN	58/72	0
19	Ethylbenzene	AlBN	68/14	0
25	Benzene	AlBN	70/24	0
28	Benzene	AlBN	68/24	0
32	neat	AlBN	70/25	0

(a) $\overline{M}_N$ = 80,000.
(b) $\overline{M}_N$ = 24,000, $\overline{M}_W$ = 50,000.

Table 2-Representative Copolymerizations of Fungicidal Monomers

Fungicidal Monomer M_1	Comonomer[a]	M_1 In Feed mole %	Initiator	System	Conditions Temp. °C/Time, hr	Yield (%)	Mole % Fungicidal Monomer in Polymer	$\lvert\eta\rvert^{30}_{THF}$ (dl/g)	$\overline{M}_N$ X10^{-4}	$\overline{M}_W$ X10^{-5}
8	n-HM	22	$K_2S_2O_8$	SLS[b]	52/5	98	15.6	1.3	8.6	50
14	n-HM	7.9	$K_2S_2O_8$	SLS[b]	48/41	34	3.2	0.9	-	-
17	n-HM	27.6	$K_2S_2O_8$	SLS[b]	70/24	51	33	0.21	1.9	3.6
10	n-HM	9.5	$K_2S_2O_8$	SLS[b]	72/22	79	9.3	1.53	-	-
13	n-BA	9.5	$K_2S_2O_8$	SLS[b]	70/18	68	4.8	13.3	9.3	40
16	MMA and n-BA	4.3	AlBN	Bz soln.	74/23	48	3.2	5.1	7.3	17
19	n-HM	11.1	$K_2S_2O_8$	SLS[b]	65/20	10	9.7	0.2	<0.5	-
19	n-HM	6.6	AlBN	Bulk	70/20	49	2.7	0.87	6.3	13.6
25	n-BMA	30	AlBN	Bz/soln.	63/28	19	1.5	0.24	3.1	8.4
28	MMA	1.1	$K_2S_2O_7$	SLS[b]	61/20	90	0	5.8	9.3	17
30	n-BA	23	AlBN	Bz/soln.	66/24	0	0	-	-	-
32	n-BA	31	AlBN	Bz/soln.	66/26	0	0	-	-	-

(a) n-HM = hexyl methacrylate, n-BA = n-butyl acrylate, n-BMA = n-butyl methacrylate, and MMA = methyl methacrylate.

(b) Emulsion polymerizations using sodium lauryl sulfate as the surface active agent.

Table 3-Sample Bulk Copolymerizations of Fungicidal Acrylates 8, 11, and 20 with Ethyl Acrylate or Vinyl Acetate[a]

Fungicidal Monomer M_1	M_2[b]	M_2/M_1	Temp. °C	Time hr	Conv. %	Mole %[c] M_1 in Copolymer	Mole %[d] M_1 in Copolymer	$\overline{M}_N$[e] X10^{-4}	$\overline{M}_W$[e] X10^{-6}	Tg[f] °C
8	EA	131	70	2.5	58	0.80	0.95	6.6	2.4	-4
8	EA	283	70	2.5	57	0.57	0.60	6.8	1.7	-6
8	EA	2,860	60	2	58	0.06	0.10	7.4	1.2	-18
8	VA	61	70	3	60	5.46	5.50	8.9	1.2	38
8	VA	329	70	3	59	1.12	1.15	7.5	1.2	35
8	VA	3,320	70	4	60	0.06	0.2	7.7	1.1	32
11	EA	40	50	3	69	1.90	2.25	9.3	3.1	1
11	EA	200	60	2.5	59	1.25	1.35	14.0	3.1	-3
11	EA	2,000	60	2.5	65	0.60	0.65	15.0	3.4	-7
11	VA	47	60	3	58	11.5	12.3	6.5	3.2	43
11	VA	233	60	3	53	2.5	2.4	5.1	2.2	37
11	VA	2,330	50	2.5	56	0.5	0.5	9.8	2.1	34
20	EA	34	60	1.5	49	1.75	1.85	8.6	3.3	-7
20	EA	267	50	2.0	70	0.61	0.70	9.4	3.2	-11
20	EA	2,670	50	2.5	71	0.48	0.65	9.3	3.0	-12
20	VA	62	50	3.0	95	1.52	1.75	6.4	3.2	37
20	VA	311	60	3.5	71	0.61	0.75	6.8	3.0	32
20	VA	3,107	60	2.0	73	0.36	0.45	7.1	2.7	30

Table 3--continued on next page

Table 3--continued

(a) Azobisisobutyronitrile was the initiator in copolymerizations of 8 while t-butylperoxypivalate was employed in compolymerizations of 11 or 20.
(b) EA = Ethyl acrylate, VA = Vinyl acetate.
(c) Determined by chlorine analysis (in copolymerizations of 8) or nitrogen analysis (in copolymerizations of 11 or 20).
(d) Calculated from pyrolysis gas chromatography determinations.
(e) Determined by GPC.
(f) Evaluated from differential scanning calorimetry (DSC) where Tg for poly(ethyl acrylate) = -24°C and poly(vinyl acetate) = 28°C.

low molecular weight, oily products.[12] These copolymerizations are summarized in Table 4 where one can see the value of Tg increases as the molar content of 21 increases.

In order to determine the reactivity of pentachlorophenyl acrylate, 8, in radical initiated copolymerizations, its relative reactivity ratios were obtained with vinyl acetate (M_2), r_1=1.44 and r_2=0.04 using 31 copolymerization experiments, and with ethyl acrylate (M_2), r_1=0.21 and r_2=0.88 using 20 experiments.[13] The composition conversion data was computer-fitted to the integrated form of the copolymer equation using the nonlinear least-squares method of Tidwell and Mortimer,[41] which had been adapted to a computerized format earlier.[42]

In this work,[13] glass transition temperatures were studied as a function of copolymer composition. Poly(pentachlorophenyl acrylate) exhibited the high Tg value of 143°. Table 5 summarizes these Tg values versus M_1/M_2 content of both vinyl acetate and ethyl acrylate copolymers.[13] The Tg values for copolymers with higher mole fractions of 8 depended on their thermal history. Tg increases after heating above Tg the first time. For example a vinyl acetate copolymer (mole fraction of 8 = 0.67) exhibited Tg values of 77°, 99°, and 103°C in three successive heating cycles.[13]

Since all the fungicidal monomers give homopolymers with a high glass transition temperature, Tg, paint binder polymers containing the fungicides must be designed to have lower Tg values. Therefore, a terpolymer acrylic binder design consisting of (1) the fungicidal acrylate, (2) methyl methacrylate, and (3) a low Tg acrylate such as n-butyl acrylate or 2-ethylhexylacrylate was selected. Several terpolymer samples were made in solution in small amounts. For example, solution terpolymerization of 16 with methyl methacrylate and n-butyl acrylate gave a good terpolymer, as described in Table 2. It is of interest that extended-chain acrylates, while resisting AIBN-initiated homopolymerization, underwent terpolymerization. This important observation cleared the way for the synthesis of stable terpolymer latices containing these slower hydrolyzing monomers. Latex formulations were needed for further mixing into sample paints for testing.

Fungicidal vinyl ethers 25, 30 and 32, resisted radical initiated homopolymerization. Copolymerizations of these monomers with n-butyl acrylate or methyl methacrylate were generally unsatisfactory. At low mole ratios of fungicidal vinyl ether to acrylate, copolymers could be isolated with either very low or no incorporation of the vinyl ethers.

Table 4-Sample Bulk Copolymerizations of 2-(4'-Thiazoyl)benzimidazole Acrylate, 21, With Ethyl Acrylate (M_2)[a]

M_1/M_2	Temp. °C	Time hr	Conv. %	Mole %[b] M_1 in Copolymer	$\overline{M}_N$[c] $\times 10^{-4}$	$\overline{M}_W$[c] $\times 10^{-6}$	Tg °C
23	60	24	48	6.4	0.7	3.0	4
49	60	24	48	4.4	1.3	2.4	4
98	55	24	53	2.5	1.6	2.1	-2
248	60	24	60	2.4	2.5	1.9	-4
498	58	8	63	1.9	2.2	1.5	-6

(a) In each run AIBN (0.04g) was used as initiator and 18 to 20g of ethyl acrylate was employed.

(b) Calculated from nitrogen analyses.

(c) Calculated from GPC.

Table 5-Glass Transition Temperatures For Pentachlorophenyl Acrylate (M_1) Copolymers With Vinyl Acetate or Ethyl Acrylate

M_2	M_1 Mole Fraction	Tg °C	M_2	M_1 Mole Fraction	Tg
--	1.00	143	--	1.00	143
VA	0.85	101	EA	0.89	83
VA	0.67	77	EA	0.73	40
VA	0.50	67	EA	0.48	18
VA	0.25	56	EA	0.02	-4
VA	0.10	48	EA	--	-24
VA	0.02	38			
VA	0.002	32			
VA	--	28			

Latex Syntheses[14]

A variety of stable latices were prepared where the latex polymer contained a bound fungicide (see Table 6).[14] In general, these were terpolymer latices using methyl methacrylate (MMA) and n-butyl acrylate (nBA) as the major acrylic monomers together with one of the fungicidal monomers. nBA was used to lower Tg. Table 6 describes 14 examples of these latices. Latices 1 and 2 are blank samples which were used as controls in accelerated growth testing. Latex 9 is a vinyl acetate system using 2-ethylhexyl acrylate as the Tg-lowering monomer.

Latices 13 and 14 (Table 6) are especially interesting because they each contain two different fungicidal monomers. Latex 13 incorporates both the acrylate, 8, and extended chain acrylate, 10, of pentachlorophenol. Thus, two different releasing groups attach the same biocide. Latex 14 incorporates the acrylates of both pentachlorophenol and tribromosalicylanilide (i.e. 8 and 14) so that it can release two different biocides.

Attempts to make acrylic latices containing o-benzyl-p-chlorophenyl acrylate, 17, using the same techniques employed for the systems in Table 6, failed.[14] It was found that no 17 was incorporated into the polymer of these latices. Similar attempts to make terpolymer latices with vinyl pentachlorophenyl ether, 25 or vinyl 3,4'5-tribromosalicylanilide ether, 28 (with MMA and nBA) failed. These fungicidal vinyl ethers were not incorporated into the resulting latex polymers. However, vinyl o-benzyl-p-chlorophenyl ether, 17, was successfully terpolymerized in latex form (see latex 12, Table 6). Formulations for latices 1-3 and 5-14 are given in Table 7.

The general procedure for the synthesis of the latices listed in Tables 6 and 7 is given below for latex 4 (Table 6).[14]

A. Preparation of Kettle Charge: Aerosol 22 (a 35% by weight aqueous solution of tetrasodium N-(1,2-dicarboxyethyl)-N-octadecylsulfonsuccinamate (10.8 g)) methanol (9.5 g), sodium bicarbonate (1.8 g) were dissolved in deionized water (176 g) and an ammonium persulfate solution (38 g of a 10% by wt aqueous solution) was added. The solution was purged with nitrogen for 15 min and charged to the polymerization vessel and heated to 60°C.

B. Preparation of Pre-Emulsified Monomer Charge: Aerosol A-102 (disodium ethoxylated alcohol half ester of sulfosuccinic acid) (10.8 g) were dissolved in water (59.1 g). The monomers were slowly added with stirring to this solution and the resulting emulsion was purged with nitrogen for 15 min. The amounts were: monomer 8 (20 g, 62 mmol), MMA (200 g, 2.2 moles) and nBA

Table 6-Composition of Stable Latices of Model Paint Binders Containing Chemically Bound Fungicides[a]

Latex Number	Monomers (mole % In Polymer): Fungicidal Monomer[b]	Comonomers[b]		Solids Content (% wt of total latex)	$\|\eta\|^{THF}_{30°}$ dl/g C	$\overline{M}_N$	$\overline{M}_W$
1	none	MMA (43)	nBA (57)	40	6.1	6.1×10^4	4.1×10^5
2	none	MMA (43)	nBA (57)	50	2.6	7.2×10^4	9.3×10^4
3	5 (5)	MMA (32)	nBA (63)	40	2.4	2.8×10^4	6.7×10^4
4	5 (2)	MMA (58)	nBA (40)	50	3.1	3.5×10^5	8.1×10^5
5	7 (6)	MMA (11)	nBA (83)	54	3.5	2.0×10^5	7.2×10^5
6	10 (6)	MMA (58)	nBA (36)	41	5.8	2.7×10^5	8.3×10^5
7	11 (2)	MMA (51)	nBA (43)	40	2.8	-	-
8	11 (6)	MMA (47)	nBA (51)	48	3.1	3.4×10^5	6.1×10^5
9	11 (1.5)	VA (87.5)	EHA (11)	54	1.3	1.2×10^5	2.8×10^5
10	13 (0.7)	MMA (52.3)	nBA (47)	42	2.7	3.5×10^5	7.7×10^5
11	16 (3)	MMA (77)	nBA (15)	37	6.4	6.3×10^5	1.6×10^6
12	23 (4)	MMA (13)	nBA (83)	43	2.0	6.6×10^5	2.8×10^5
13	5 (8) & 7 (8)		nBA (84)	48	1.3[d]	e	e
14	5 (9) & 11 (9)		nBA (82)	37	1.4[d]	e	e

(a) Latices were prepared as described in experimental section of the text. MMA = methyl methacrylate, nBA = n-butyl acrylate, VA = vinyl acetate, EHA = 2-ethylhexyl acrylate.

(b) The numbers in the brackets refer to the mole % of the monomers in the binder polymer. The binder polymer was isolated from the coagulated latices and subsequently purified prior to elemental analyses.

(c) Intrinsic viscosities are those of the binder polymer after coagulation, isolation and purification by two reprecipitation steps.

(d) $|\eta|$ values in DMF at 30°C.

(e) Molecular weight measurements by gel permeation chromatography were not possible since these polymers were insoluble in THF and $CHCl_3$.

Table 7-Specific Formulations for Latices 1-3 and 5-14 (Table 3)
(All Weights are in Grams)

	Latex No.						
	1	2	3	5	6	7	8
Kettle Charge							
Aerosol 22	13.4	15	12.9	2.6	3.4	4	18.3
$(NH_4)_2S_2O_8$	1.9	1.8	1.8	0.4	0.4	1.1	3.7
CH_3OH	5.8	6	5.6	1.2	1.2	2.4	9.5
H_2O	138	133	134	26.6	27.9	54	211
$NaHCO_3$	1.6	1.2	1.3	0.2	0.3	0.8	1.8
Monomer Charge							
MMA	100	95	100	20.8	18.8	13.4	220
nBA	90	89	71.5	17.8	14.7	39	160
Fungicidal monomer	-	-	11	1.6	5.1	16.6	20.2
H_2O	78	27	88	7.3	18.5	32.1	59
Aerosol 102	16.6	16.5	16.5	3.3	3.5	4	17.9
Sodium metabisulfate	0.5	0.5	0.6	0.1	0.1	0.4	0.8
CH_3OH	6.0		6.0	1.1	1.3	3.8	9.7
Delayed Charge							
Itaconic acid	2.5	2.5	2.5	0.5	-	1.1	3.7
$(NH_4)_2PO_4$	0.3	0.3	0.3	0.1	-	0.2	0.6
H_2O						5.3	19.3

	Latex No.					
	9	10	11	12	13	14
Kettle Charge						
Aerosol 22	11.6	3.7	1.7	3.4	4	5
$(NH_4)_2S_2O_8$	3.4	0.4	0.3	0.6	-	-
CH_3OH	9.5	1.3	0.6	1.3	-	-
H_2O	206	30.3	13.9	27.8	6	50
$NaHCO_3$	3.4	0.3	0.1	0.3	0.8	0.8
$K_2S_2O_8$	-	-	-	-	0.7	0.7
Monomer Charge						
MMA	-	18.8	10.1	20	-	-
nBA	-	14.3	7.3	35	62.1	57.8
VA	322.2	-	-	-	-	-
EHA	60	-	-	-	-	-
Fungicidal monomer	20	2.6	1.1	2.2	19.2 (5) 21.6 (7)	16 (5) 25 (11)
H_2O	59.4	19.8	20	18.7	20	40
Aerosol 102	11.1	3.8	1.9	3.6	6	5
Sodium metabisulfite	0.8	0.2	0.1	0.2	0.3	0.3
CH_3OH	9.6	1.4	0.8	1.3	5	-
Benzene	-	-	-	-	-	50
Delayed Charge						
Itaconic acid	3.8	-	-	-	-	-
$(NH_4)_2PO_4$	0.6	-	-	-	1	-
H_2O	19.7	-	-	-	20	-

(160 g, 1.25 moles). The resulting monomer emulsion was charged to reaction vessel (A) from a dropping funnel fitted with a stirrer and agitation was maintained in the dropping funnel during the entire addition procedure.

C. Preparation of Delayed Charge: A delayed charge of itaconic acid (3.8 g), diammonium phosphate (0.6 g) in water (19.1 g) was purged with nitrogen for 15 min.

D. Polymerization: After the reaction kettle (with its initial charge) reached 60°C, the preemulsified monomer charge was added at a rate of 1-2% per minute. When approximately 15% of the monomer charged had been added, the addition was stopped until polymerization was indicated by an exotherm. The delayed charge was then added to the kettle and the monomer charge was started again. After all the monomer had been added, the polymerization was continued for 1 hr at 60°C, cooled and filtered through a 100 mesh nylon screen to remove any coagulated material. A portion of this latex was poured into a 7:1 methanol/saturated aqueous NaCl solution to precipitate the polymer which then was twice reprecipitated from THF into methanol. Analysis: C 61.09, H 8.14, Cl 2.95.

Derivatization of Preformed Polymers with Fungicidal Moieties

Two techniques were available to bind biocides to polymers. The first, involving synthesis of a biocidal group as part of a monomer followed by polymerization, has been described above. The second method involved direct reaction of the fungicide with the preformed coating polymer containing a suitable functional group.[12,13] For example, a vinyl acetate-acrylic acid (95:5) copolymer was prepared and the carboxylic acid units were converted to acid chloride groups with excess thionyl chloride. Reaction of the polymeric acid chlorides with the -OH or ≥NH groups of the fungicides could lead to an anchored fungicide. This technique, however, was less than satisfactory. The polymeric acid chlorides were very sensitive to hydrolysis and polymer degradation occurred, presumably due to the presence of HCl. This decomposition was reflected by a decrease in viscosity during this reaction sequence. Furthermore, the final polymers exhibited strong, broad hydroxyl stretching bands in their IR spectra. Transesterification was also attempted unsuccessfully. Poly(ethyl acrylate) was reacted with pentachlorophenol in the presence of p-toluene sulfonic acid in benzene. Very little ethanol was isolated and upon precipitation of the polymer in petroleum ether only oily material remained.

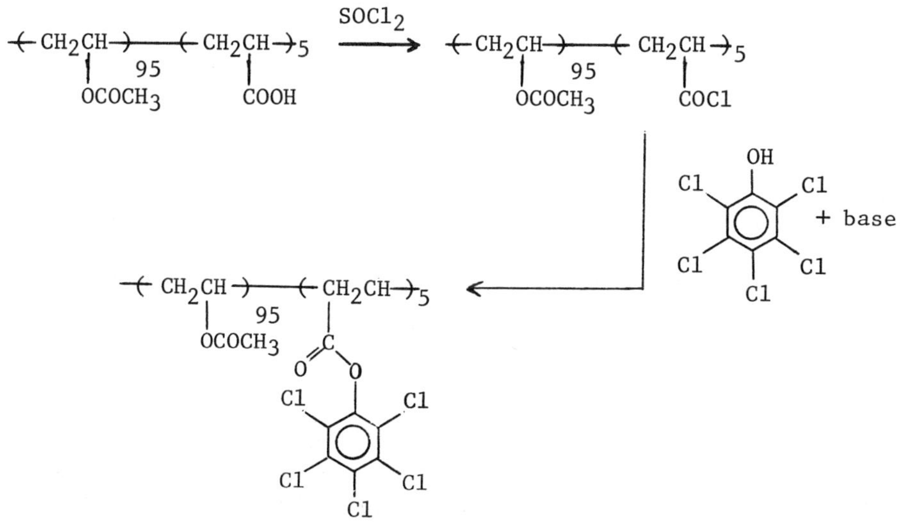

Evaluations of the Biological Activity of Polymers With Chemically Bound Biocides

The question of whether or not the polymer-anchored fungicides described earlier can be active against microorganisms can be subdivided into several specific questions. Do these materials exhibit any activity? Are they more or less active than the unanchored biocide when compared at equal concentrations of active agent? Are they active under certain conditions and inactive under other conditions where their unanchored analogs are active? Must the biocide be released by the polymer in order to be active or is the macromolecular form also active? If the biocide must be released from the polymer, how is activity effected by the rate of release? How is the rate of release dependent upon the functional group connecting the agent to the polymer and upon the polarity and structure of the polymer? Finally, can a suitable polymer-anchored fungicide system be designed to function in a paint to confer a longer mildew resistant period on surfaces coated with that paint?

Minimum Inhibitory Concentration Tests

The question of whether or not a polymer-anchored biocide can be active against microorganism growth was first tested by comparing the minimum inhibitory concentrations (MIC) of pentachlorophenol, its acrylate, 8, and the homopolymer of pentachlorophenyl acrylate 8. An agar dilution method was used. To a sterilized agar, a solution of the compound to be tested was added in the appropriate amount to make final concentrations of 1000, 500, 250, 100, 50, and 10 ppm. Following inoculation, the agar dishes were incubated at 30° for seven days. The results are summarized in Table 8. It can be seen that pentachlorophenol's MIC is lower than that for acrylate 8 for all the test organisms except E. Coli. However, 8 is a good biocide. The homopolymer of 8 exhibits MIC values the same as that of 8 against Staphylococcus aureus, Aspergillus niger, and Penicillium citrinum and a slightly higher value against Saccharomyces cerevisiae. Despite good control of these four organisms, the homopolymer was not effective in control of E. Coli. It is clear the homopolymer does function as a biocide but with a different specificity from either pentachlorophenol or 8. In identical tests, free 2-thiazoylbenzimidazole, 6, exhibited MIC values (in ppm) of 250 for E. Coli, 1000 for Staphylococcus aureus, 50 for Aspergillus niger, 250 for Penicillium citrinum and 100 for Saccharomyces cerevisiae.

Acrylate 8 and its homopolymers were also compared against Aspergillus niger using a paper disk/agar dish method. The biocidal compound was added to a polyvinyl acetate emulsion which was coated onto filter papers. The amount of coating doubled the weight of the filter paper after drying. The coated filter papers were placed in the center of an agar dish, inoculated, and incubated at 30° for 28 days. Acrylate 8 prevented growth at 3000 ppm versus about 7000 ppm for its homopolymer via this method.

The MIC values of several fungicidal monomers and polymers are summarized in Table 9 against Pseudomonas sp., Aureobasidium pullulans and Alternaria sp. These studies were performed in water nutrient broth solutions by standard procedures. First, it is interesting to compare pentachlorophenol with its acrylate, 8, and its chain-extended acrylate, 10. All three are effective against Aureobasidium pullulans and 10 while quite active, is generally less active than pentachlorophenol or 8. When ester hydrolysis of 10 occurs, the resulting product is alcohol 9. It may or may not be the active species. Further reaction of 9 to give pentachlorophenol may not be occuring in the nutrient medium.

Acryloyloxy-3,4',5-tribromosalicylanilide, 14, was similar in activity to pentachlorophenol against all four organisms tested, whereas o-benzyl-p-chlorophenol acrylate is less active against Alternaria sp. and Aureobasidium pullulans and more active against cladosporium sp. These differences in activity suggest that the use of two or more biocides in coatings might prove useful under some conditions. The chain-extended acrylate, 19, is less active in all the tests, suggesting its hydrolysis product, 18, may not be as active as 17 or o-benzyl-p-chlorophenol.

Table 8-Minimum Inhibitory Concentrations[a]
Of Pentachlorophenol, Pentachlorophenyl Acrylate, 1, And Its Homopolymer
Toward Microorganisms Using An Agar Dilution Method

Compound	Microorganism	MIC (in ppm)					
		1000	500	250	100	50	10
Pentachlorophenol	E. Coli	-	-	-	-	+	+
	Staphylococcus aureus	-	-	-	-	-	-
	Asperigillus niger	-	-	-	-	+	+
	Penicillium citrinum	-	-	-	-	-	+
	Saccharomyces cerevisiae	-	-	-	-	-	+
1	E. Coli	-	-	-	-	+	+
	Staphylococcus aureus	-	-	-	-	+	+
	Aspergillus niger	-	-	-	+	+	+
	Penicillium citrinum	-	-	-	+	+	+
	Saccharomyces cerevisiae	-	-	-	-	+	+
Homopolymer of 1[b]	E. Coli	+	+	+	+	+	+
	Staphylococcus aureus	-	-	-	-	+	+
	Aspergillus niger	-	-	-	+	+	+
	Penicillium citrinum	-	-	-	+	+	+
	Saccharomyces cerevisiae	-	-	-	+	+	+

(a) Determined by the agar dilution method.
(b) M_N = 10,800, M_W = 24,600.

Table 9-Minimum Inhibitory Concentrations (in PPM) of Selected Fungicidal Monomers and Polymers

Monomer or Polymer (mole % polymer composition)	Pseudomonas sp.	Cladosporium sp.	Aureobasidium pullulans	Alternaria sp.
Pentachlorophenol[a]	300-500	100-250	<50	50-100
Pentachlorophenyl Acrylate, 1	300-500	100-250	<50	100-250
2-Pentachlorophenoxyethyl Acrylate, 5	1000	250-500	100	500-1000
Acryloyloxy-3,4',5-Tribromosalicylanilide, 3	300-1000	100-250	50	50-100
o-Benzyl-p-chlorophenyl Acrylate, 4	250-500	50	250-500	1000
2-(o-Benzyl-p-chlorophenoxyl)ethyl Acrylate, 8	500-1000	500-1000	1000	1000
Poly(1/MMA/nBA)[b,c] (5/32/63)	500-1000	250-500	100-250	250-500
Poly(5/MMA/nBA)[c] (6.1/11.4/82.5)	500-1000	500-1000	100-250	250-500
Poly(5/HMA)[c] (9.3/90.7)	1000	500-1000	1000	1000
Poly(4/HMA)[c] (32.6/67.4)	1000	230-460	-[d]	-[d]
Poly(8/HMA)[c] (9.7/93)	500-1000	180-360	-[d]	-[d]

(a) MIC for Trichophyton mentagraphtes was ~500 ppm.

(b) Latex terpolymer, $|\eta|^{30}_{THF}$ = 3.1 dl/g.

(c) MIC for polymer samples are calculated based upon the weight of biocidal monomer in the polymer which was added to the broath.

(d) Only tested up to 200 ppm (of biocide) where no activity was observed.

Of greatest interest are the terpolymers and copolymers containing 8, 17, 10 and 19. These polymers form a separate microdispersed phase in stirred broth solutions since the polymers are not water soluble. MIC values were calculated based on the weight of biocide added and not on the total weight of polymer added. The MMA/nBA terpolymer of 8 was active in control of microorganism growth. Copolymers and terpolymers of 17 and 19 showed little activity against Aureobasidium pullulans or Alternaria sp., but the polymers were only tested in amounts below an equivalent of 200 ppm of biocidal monomer (bound to polymer) in the medium. At this level neither of the monomers 17 or 19 were active. Remarkably, the hexylmethacrylate copolymers of 17 and 19 exhibited MIC of 230-460 and 180-360 ppm, respectively, against Cladosporium sp. Thus, the MIC of free 17 was only 5 to 10 times less than that of its polymer while free 19 was somewhat less active than its polymer.

The results of experiments of the type summarized in Table 8 suggests that polymer-bound biocides can be active under a variety of conditions. As expected, the polymeric systems are usually not as active as the free fungicide, presumably either because the concentration of free biocide is lower in the polymeric cases or because the inherent activity of the macromolecular biocide is less than the free biocide.

Agar Dish Accelerated Growth Tests

Agar dish tests were also carried out. In such tests, a tile square or filter paper squares were coated with polymer-anchored fungicides, controls, or polymers where the free biocide was blended. After inoculation and incubation under favorable growth conditions, the growth pattern was examined. Figure 1 illustrates three situations. The dish on the left illustrates a control sample where no biocide is present. Mildew growth has overrun the square sample in the center of the dish. The center dish contains a sample protected by a monomeric fungicide which has been blended into the polymer which coats the square sample. The square sample is free of mildew growth due to the presence of the fungicide. Furthermore, notice the large zone of growth inhibition which extends out from the square into the surrounding agar medium. This zone results from the migration of the free fungicide out of the polymer into the agar. In this case, the rate of migration is fast enough to result in a biologically significant concentration of the biocide in the surrounding agar, hence the resulting growth inhibition. Further from the square, mildew growth is seen in the agar.

In vitro agar dish tests of 8-hydroxyquinolyl acrylate, 11, and its copolymers with ethylacrylate and vinylacetate against Aureobasidium pullulans were performed using nutrient media of

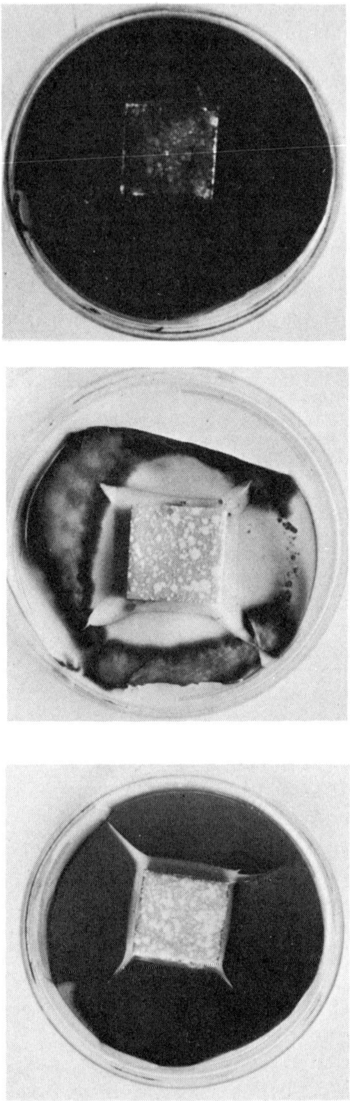

Figure 1. Representative agar dish growth tests of A. pullulans *and* Alternaria *on polymer films deposited on ceramic tiles. Key: top, unprotected film—mildew growth over entire dish; middle, film protected by fungicide-polymer film—zone of inhibition around tile; and bottom, film protected by polymer-bound fungicide—film is protected without wasteful zone of inhibition.*

malt extract agar or yeast extract-malt extract-glucose agar. One-inch squares of sterile Whatman #1 filter paper were immersed in solutions of 11 or its polymers, then dried, and the square was placed in contact with the agar. Inoculation was performed either by pouring an upper layer of inoculated agar into the dish or by using two circular plugs (one cm diameter) of an actively growing petri dish culture of Aureobasidium pullulans which were shaken with the nutrient broth prior to its addition to the dishes. The results, summarized in Table 10 show that heavy growth up to the edge of the squares occurred using monomer 11. After two days, a very small zone of inhibition was observed about the square. These zones grew quite large in 10 days, showing that biocide 2 was active and that it migrated from the filter paper. The copolymer of 11 and ethyl acrylate (with two mole percent of 11) initially showed zones of inhibition which decreased with time in the low concentration regime as the concentration gradient of biocide decreased with time. Quite clearly, all the copolymers of 11 show a similar behavior. Hydrolysis of 8-hydroxyquinoline from the polymer is fast and leaching leads to a zone of inhibition which is rapidly established. Slowly, Aureobasidium pullulans invade this area as less and less biocide remains. These studies show that polymers of 11 are not suitable for coating use because hydrolytic cleavage of the active agent occurs too rapidly for long retention in the coating. This rapid hydrolysis is due to neighboring group participation (Scheme VII).

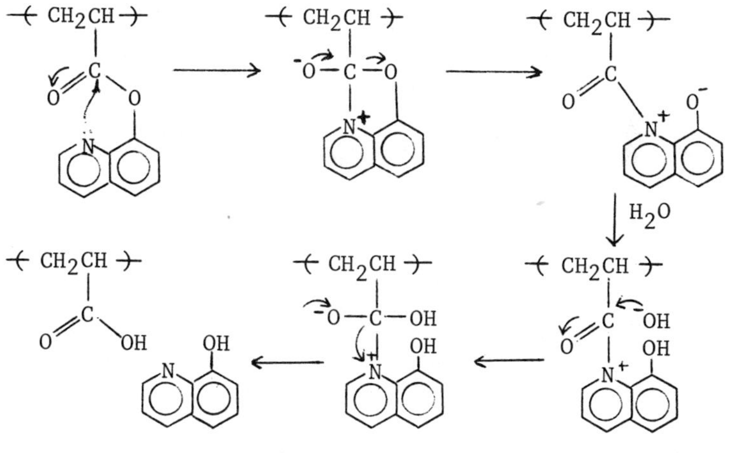

SCHEME VII - Self-Catalysis Mechanism

Similar in vitro studies were also conducted on copolymers of 8 and 14 as well as on three terpolymers containing 14 (See Table 11). Both free pentachlorophenol and monomer 14 give zones of inhibition in control experiments because they migrate from

Table 10–In Vitro Inhibition of Aureobasidium Pullulans By Ethyl Acrylate (EA) and Vinyl Acetate (VA) Copolymers Of 8-Hydroxyquinolinyl Acrylate, 11[a]

Biocide (Copolymer Composition) in mole %	Time (in Days)	Concentrations of Biocide, % W/V 1.0		5.0	
		ME[b]	YM[c]	ME[b]	YM[c]
8-Hydroxyquinolinyl Acrylate, 11	1	H	H	H	H
	2	tr	tr	tr	tr
	10	25	25	25	25
Copolymer of 11 and EA (2/98)	1	3	3	12	12
	2	6	6	12	12
	10	tr*	H*	12	18
Copolymer of 11 and EA (0.6/99.4)	1	H	H	H	H
	2	tr	tr	tr	tr
	10	H*	H*	H*	H*
Copolymer of 11 and VA (13.4/86.6)	1	H	H	6	9
	2	tr	tr	9	12
	10	H*	H*	6*	M*
Copolymer of 11 and VA (5.4/94.6)	1	tr	tr	6	3
	2	tr	tr	6	6
	10	M*	M*	6	tr*
Copolymer of 11 and VA (0.5/99.5)	1	H	H	H	H
	10	H	H	H	H

(a) Designations in the table: H = heavy growth on agar up to the edges of the filter paper impregnated with biocide (but no growth on that filter paper; M = moderate growth up to edge of filter paper; tr = minimal zone of inhibition about the edge of the filter paper. Arabic numbers indicate the diameter (in mm) of a zone of growth inhibition surrounding the filter paper. * indicates the original zone of inhibition was being reinvaded by Aureobasidium pullulans growth.

(b) ME = malt extract agar.

(c) YM = yeast extract - malt extract - glucose agar.

Table 11-In Vitro Inhibition of Aureobasidium Pullulans By Copolymers and Terpolymers Containing Chemically Attached Fungicides, Effect of UV Irradiation

Biocide[a] (Copolymers mole % Composition)	Concentration % (W/V)	Size of Zone of Inhibition (in cm) on Days: 2	4	8	16	30
Pentachlorophenol	2	1.3	0.96	0.81	-	0
Copolymer of 8 and HMA (15.6/84.4)	0.5	0	0	0	0	0
	2	0	0	0	0	tr
	2 (UV)[b]	0.6	1.3	2.8	3.2	3.2
Acrylolyloxy-3,4',5-tribromosalicylanilide, 14	2	1.6	1.8	2.0	-	Complete
Copolymer of 14 and HMA (3.2/96.8)	0.5	0	0	0	0	tr
	2	0	0	0	0	tr
	2 (UV)[b]	0.6	1.3	2.8	3.2	3.2
Terpolymer of 14, MMA, nBA (2/47/51)	2	0	0.4	0.8	1.5	1.8
	2 (UV)[b]	0.3	0.7	1.2	1.8	Complete
Terpolymer of 14, MMA, nBA (2/58/40)	2	0	0.2	0.7	1.3	1.7
	2 (UV)[b]	0.2	0.5	0.8	1.6	Complete
Terpolymer of 14, VA, EHA	2	0	0.5	1.0	1.8	1.9
	2 (UV)[b]	0.2	0.6	1.1	1.9	Complete

(a) HMA = Hexyl methacrylate, MMA = methyl methacrylate, nBA = n-butyl acrylate, VA = vinyl acetate, EHA = 2-ethylhexyl acrylate. Medium used was malt extract agar.

(b) Impregnated filter paper was irradiated with a UV lamp 32 cm above sample for 164 hr prior to addition to petri dish.

the impregnated paper squares into the agar. Copolymers of 8 or 14 with hexyl methacrylate do not produce zones of inhibition. All three terpolymers of 14 however, began to give such zones after four days, and the size of the zones increased slowly up to 30 days. This shows that slow release of 3,4',5-tribromosalicylanilide occurs from the terpolymers under the moist test conditions. The presence of MMA or vinyl acetate makes the terpolymers more polar than the HMA copolymers. This may facilitate ester hydrolysis to release the fungicide. This difference in behavior suggests that biocide release rates could be influenced by the polymers' composition as well as the functional group used to attach it. When the impregnated samples were exposed to UV irradiation prior to inoculation, zones of inhibition were always observed in two days. Irradiation causes photochemical ester cleavage. Thus, pentachlorophenol and 3,4',5-tribromosalicylanilide are cleaved from the polymer and they rapidly migrate into the surrounding agar creating a zone of inhibition.

Identical in vitro studies of terpolymers of methyl methacrylate/n-butylacrylate/and acrylates 10, 13, 16 and 19 were conducted and example results are summarized in Table 12. None of these terpolymers gave a zone of inhibition during the 30-day test. Thus, the chain-extended acrylate polymers resist hydrolysis better than their non-extended analogs (Table 11) as was expected based on relative leaving group abilities. All four polymers listed in Table 12 prevented growth on the impregnated filter paper when compared to a MMA/nBA copolymer blank. The terpolymer containing 10 continued to prevent any growth for 30 days, whereas light growth was observed in 30 days for terpolymers with 13 and 16. Light growth was observed on the squares after eight days with the terpolymers of 19.

Effect of UV Exposure

Exposure of these polymers to UV irradiation (160 hr) prior to inoculation led to zones of inhibition. As shown in Table 12, initial zones of growth inhibition extended into the agar, indicating that photohydrolytic cleavage of a portion of the bound fungicide occurred. These initial zones of inhibition decreased with time as this released fungicide was consumed. The impregnated filter paper squares continued to be protected as in the case of the unirradiated samples. It is not known if UV irradiation frees the parent fungicides via photochemical ether cleavage of if the fungicidal 2-hydroxyethyl derivatives are released by photochemical ester hydrolysis.

The *in vitro* tests of UV-irradiated samples summarized in Tables 12 and 13 emphasize an important point. Photochemical cleavage must be considered in the design of biocidal coatings. UV may promote a faster biocide release than is desirable.

Table 12-In Vitro Inhibition of Aureobasidium Pullulans By Fungicides Bound To Polymers, Extended-Chain Acrylate Ester Functions[a]

Biocide Terpolymer[c] (Mole % Composition)		Growth Response and Size of Zone of Inhibition (in cm.) on days:[b]			
		2	4	8	30
Terpolymer of 10, MMA, nBA		0(-)	0(-)	0(-)	0(-)
(3.3/42.4/54.3)	(UV)	0.2(-)	0.2(-)	0(-)	0(-)
Terpolymer of 13, MMA, nBA		0(-)	0(-)	0(-)	0(+)
(6.3/57.7/36.3)	(UV)	0.5(-)	0(-)	0(-)	0(+)
Terpolymer of 16, MMA, nBA		0(-)	0(-)	0(-)	0(+)
(0.7/MMA, nBA)	(UV)	0.3(-)	0.4(-)	0.1(-)	0(+)
Terpolymer of 19, MMA, nBA		0(-)	0(-)	0(+)	0(+)
(3/77.4/19.6)	(UV)	0.1(-)	0(-)	0(-)	0(+)
Control Copolymer MMA, nBA (60/40)		0(+)	0(+)	0(++)	0(+++)

(a) Yeast extract-malt extract-glucose agar used.
(b) Zone of growth inhibition around the filter paper given in arabic numbers. (-) means no growth on the filter paper (+) means light growth on filter paper (++) moderate growth on filter paper and (+++) heavy growth on filter paper.
(c) 10 = 2-pentachlorophenoxyethyl acrylate, 13 = 2-(8-quinolinyloxy)ethyl acrylate, 16 = 2-(2-acryloylethoxy)-3,4',5-tribromosalicylanilide, 19 = 2-(o-benzyl-p-chlorophenoxy)ethyl acrylate.

Table 13-Relative Mildew Growth On Polymer And Paint Samples Exposed To Aureobasidium Pullulans for Two Months By The Sapwood-Substrate Enriched Vermiculite Method (Zabel Test)

Coating	Visual Ranking[a]		Stain Intensity on Wood Substrate	
	Pine	Basewood	Pine	Basewood
Unweathered				
PRI-Reference Paint	3.7	2.7	severe	medium
MMA/nBA Latex[b]	1.7	1.3	severe	medium
MMA/nBA Latex blended with 8/MMA/nBA Latex[c]	1.0	0.7	severe	medium
MMA/nBA Latex + 1% pentachlorophenol[b]	1.0	0.7	medium	light
MMA/nBA Latex + 2% pentachlorophenol[b]	0.7	0.7	light	light
MMA/nBA Latex with 10/MMA/nBA Latex[d]	1.7	1.7	severe	medium
Weathered				
PRI-Reference Paint	3	3	medium	medium
MMA/nBA Latex blended with 8/MMA/nBA Latex[c]	0.7	0.7	medium	medium
MMA/nBA Latex + 1% pentachlorophenol[b]	0.3	0	light	light
MMA/nBA Latex blended with 10/MMA/nBA Latex[d]	0.7	1.3	medium	medium

(a) Visual rating scale 0 = no growth, 1 = trace, 2 = light, 3 = moderate and 4 = heavy. Observed with 20X binocular microscope.

(b) Copolymer latex of methyl methacrylate (MMA), 42.8 mole % and butyl acrylate (nBA), 57.2 mole %. $|\eta|^{30°}_{THF}$ = 2.6 dl/g (on purified polymer), latex = 50% solids.

(c) 7.6 Grams of the MMA/nBA latex described in footnote b was blended with 10 g. of a terpolymer latex of monomer 8 (5 mole %), MMA (63.4 mole %), and nBA (31.7 mole %), 40% solids.

(d) 12.1 Grams of the MMA/nBA latex described in footnote b was blended with 5 g. of a terpolymer latex of monomer 10 (6.1 mole %), MMA (11.4 mole %), and nBA (82.5 mole %), $|\eta|^{30°}_{THF}$ = 2.3 dl/g, latex = 54% solids.

"Zabel" Accelerated Growth Tests

The relative mildew growth on polymer and on paint samples exposed to Aurobasidium pullulans was also studied using the "Zabel Test."[43] In this procedure, veneer strips of southern yellow pine and bass wood are sterilized, cooled, and painted (2 coats with 24 hrs. drying time). After 2-3 days the strips were soaked in distilled water, sterilized at 100°C and placed vertically in 16-ounce French square bottles containing 20 gm. of vermiculite. A micronutrient enriched nutrient media was then poured into the vermiculite (100 ml per bottle). After sterilization at 121° the painted strips were positioned so that the bottom of the painted zone was 0.5 cm above the vermiculite surface. The specimens were inoculated with a spore suspension of Aureobasidium pullulans (16624) by pipette, incubated two months at 28°, and then rated visually with a 20x binocular. In the weathered series, the specimens were exposed four days in an Atlas Weather O Meter (UV and water spray). Representative results are summarized in Table 13.

A standard Paint Research Institute latex paint, provided by J. F. Matthews of DuPont, was the reference sample to which the biocidal latex polymers were compared. After two months both the unweathered and weathered samples showed moderate or moderate to heavy growth on the paint surface. The unweathered samples showed severe stain intensities on basswood. The weathered samples showed medium stain intensities on both. When coatings of an MMA/nBA latex blended with a MMA/nBA/8 terpolymer latex were tested, only trace growth of Aureobasidium pullulans on the coatings was observed using pine or basswood substrates. A portion of this control can be attributed to the presence of the anchored pentachlorophenol. It should be noted that the control coatings prepared from an MMA/nBA (43/57) latex supported less growth than the PRI reference paint but more growth than the coating prepared from the MMA/nBA/8 terpolymer latex. The protection given by this terpolymer latex was equivalent to that given by the blending of 1% pentachlorophenol into the MA/nBA latex. However, under these test conditions, the terpolymer latex of MMA/nBA/10 did not reduce growth below the level occurring on the MMA/nBA latex coatings.

The stain intensity on the wood under the coatings decreased markedly when pentachlorophenol was blended. Since the biocide was not attached to the polymer, it migrates into the wood thereby retarding growth. No reduction of stain intensity occurred using the polymer-attached biocide latices. This shows migration of biocide out of the coating into the wood does not occur in amounts that are biologically significant.

Conclusions

The combined evidence from accelerated growth testing, including the results summarized here, suggest that polymer-bound biocides are worthy of further study. Clearly, the accelerated testing described herein is no substitute for fence exposures of actual test paints which have been formulated from latices of the terpolymers (MMA, nBA, fungicidal monomer type). But, it does appear clear that polymers containing chemically bound biocides can be biocidal to a variety of microorganisms. A preliminary conclusion is that polymers containing the fungicidal acrylates are usually less active than the free parent fungicide but more active than the chain-extended acrylates. Little data is available for the vinyl ether polymers.

It is also clear that the effect of UV irradiation must be considered. Photochemical cleavage can cleave the biocide under conditions where the ester linkage to the polymer would otherwise be stable. Whether this is a blessing or a curse depends upon the system design, but it must be considered. Changing the functional group linkage or the biocide will alter the importance of photochemical cleavage. The polarity of the polymer will also play a role in the rate of biocide release. Hydrolytic cleavage proceeds by an ionic mechanism and is sensitive to medium effects.

All of the biocidal monomers should be formulated into latices and these, in turn, made into test paints for fence exposure studies. It would be a poor strategy to just test single monomer this way. The failure of a single system would not mean that many other biocidal polymer paint binders might not work successfully. A variety should be tested.

During the course of these studies, a large number of copolymer, terpolymer, and latex samples containing chemically bound biocides were given to the Mildew Steering Committee. Other samples were sent to William Singer (Troy Chemical Co.) and a series of latices were sent to Jerry Noren (DeSoto Inc.). In most cases, biological testing on these systems, or paints formulated from them, has not yet been carried out. Since only limited quantities were available (less than 400 g for all latices), test fence studies have not been conducted. Under the direction of Professor John Schmitt a number of copolymer coatings were exposed on fences in South San Francisco, CA and Brownsville, TX. Mycological sampling followed the successional ecology. Unfortunately, support of this work ended before the most interesting terpolymer latices of fungicidal acrylates, extended acrylates, and vinyl ethers could be studied.

Acknowledgment

We gratefully recognize the cooperation of several research groups in performing portions of the experimental studies summarized in this paper. They include the groups of Professor Robert Zabel (Department of Environmental and Forest Biology, State University of New York, Syracuse), Professor John Schmitt (Department of Botany, Ohio State University, Columbus) and Dr. Opperman (Cosan Chemical Co.). Details of their work may appear elsewhere. This research was supported, in part, by a Paint Research Institute Mildew Defacement Grant, No. MI-57 at the University of Alabama. A large fraction of the monomer synthetic work, polymer preparation, and latex formulations described here was done by Mr. K. Lawyer and may be found in his MS Thesis, University of Alabama 1980.

Literature Cited

1. Pittman, C. U. Jr. J. Coatings Tech., 48 (617), 31 (1976).
2. Steele, M. D. and Drisko, R. W., ibid, 48 (616), 59 (1976).
3. Steele, M. D. and Drisko, R. W., Technical Note N-1439, Civil Engineering Laboratory, Naval Construction Battalion Center, Port Hueneme, CA, June 1976.
4. Drisko, R. W., O'Neill, T. B., and Schwaub, L. K., Technical Note N-14-18, Civil Engineering Laboratory, Naval Construction Battalion Center, Port Hueneme, CA, May 1977.
5. O'Neill, T. B. and Drisko, R. W., Journal of Coatings Technology, 50, No. 641, 31 (1978).
6. Brand, B. G. and Kemp, H. T., "Mildew Defacement of Organic Coatings," Federation of Societies for Coating Technology, Philadelphia, PA (1978).
7. Schmitt, J. A., "Growth Patterns of Paint Mildew and Aureobasidium pullulans," Journal of Coatings Technology, 50, No. 639, 51-53 (1976).
8. Ludwig, L. E., Journal of Paint Technology, 46, No. 594, 31 (1974).
9. Stahl, G. A., and Pittman, Jr. C. U., Journal of Coatings Technology, 50, No. 639, 62 (1978).
10. Paul, D. R. and Harris, F. W., ACS Symposium Series, 33, 1976.
11. Scher, H. B., ACS Symposium, 53, 1977.
12. Pittman, Jr., C. U., Stahl, G. A., and Winters, H., Journal of Coatings Technology, 50, No. 636, 49 (1978).
13. Pittman, Jr., C. U. and Stahl, G. A., J. Appl. Polym. Sci, 26 (7), 2403 (1981).
14. Pittman, Jr., C. U., Ramachandran, K. S., and Lawyer, K. R., J. Coatings Technol., 54 (690), 27 July 1982.
15. Pittman, Jr., C. U. and Lawyer, K. R., ibid, 54 (690),41 July 1982.

16. Pittman, Jr., C. U., Lawyer, K. R., and Ramachandran, K. S., in ACS Symposium Series No. 186, Biological Activities of Polymers, C. E. Carraher and C. G. Gebelein, editors, 1982, pp 35-53.
17. Leclerq, A., Bull-Rech. Argon. Gembloux, 12, 215 (1977).
18. Chatterjee, M. G., Ranganathan, S. K., Saxene, B. L., and Sengupta, S. R., Defence Sci. J. (New Delhi), 11, 170 (1961).
19. Ismail, R., French Pat. 1,547,223, Nov. 22 (1968).
20. Cirelli, D. P. Federal Register, 48, 48446 (1978).
21. Albert, A. and Magrath, D., Biochem. J., 41, 534 (1947).
22. Phillips, J. P., Chem. Rev., 56 271 (1956).
23. Gershon, H. and Parmegiani, R., Appl. Microbiology, 10, 348 (1962).
24. Lemaire, H., Schramm, C. H., and Cahn, A., J. Pharmaceutical Sci., 50, 831 (1961).
25. Jungermann, E., Brown, Jr., J., Yackovitch, F., and Taber, D., Journal of American Oil Chemists Society, 44, 232 (1967).
26. Monar, M. N., J. Am. Oil Chemists' Soc., 45, 729 (1968).
27. Monsanto Industrial Chemicals Co., Technical Bulletin IC/DP-2, February 1977.
28. Klarmann, E., Gates, L. W., and Shternov, V. A., J. Am. Chem. Soc., 54, 3315 (1932).
29. Adelman, R. L., J. Chem. Soc., 75, 2678 (1953).
30. Foster, D. J. and Tober, E., J. Am. Chem. Soc., 83, 851 (1961).
31. Cristau, H. J. and Soleiman, M., Synthesis, 736 (1936).
32. I. G. Farbeinindustrie, British Pat. 443,978, March 11, 1936; CA. 30; p52376.
33. Skortsova, G. G., Andriyanko, M. A., and Mansurov, Y. A., Khim-Farm. Zh., 10, 22 (1976); CA, 86: 5288k.
34. Skortsova, G. G., Tyrina, S. M., Panchenkov, A. F., Shamis, L. A., and Vornov, V. K., Zh. Obsch. Khim., 40, 2039 (1970); CA 74: 71047j.
35. Nakachi, O. and Yasuda, M., personal communication.
36. Trostina, V. A., Khokhlora, F. A., and Blagova, S. N., Plast. Massy. 56 (1977).
37. Felton, L. C. and McLaughlin, C. B., J. Org. Chem., 12, 298 (1947).
38. Klahre, G., Hubele, A., and Huber-Emden, H., Swiss Pat. 555,139, 31 October 1974, CA 82: p81694x.
39. Nishimo, Jr., Yokota, T., Fujii, S. and Sogen, K., Kobunski Ronbunshu, 31, 298 (1974); CA 83: 59348m.
40. Carlson, W. W., U. S. Patent 2,448,767, September 7, 1949.
41. Tidwell, P. W. and Mortimer, G. A., J. Polym. Sci. Part A-1, 3, 369 (1965).
42. Pittman, Jr., C. U. and Rounsefell, T. D., in Computers in Chemistry and Instrumentation, Vol. 6, J. S. Mattson, H. C. MacDonald, Jr., and H. B. Mark, Eds., Marcel Decker, New York, 1977.
43. Zabel, R. A., and Horner, W. E., J. Coatings Technol., 53, (625), 33 (1981).

RECEIVED June 13, 1983

BULK EFFECTS OF ENVIRONMENTS

9

Typical Uses and Durability of Various Plastics in Hostile Environments

A. A. BOOVA

Atlas Minerals & Chemicals, Inc., Farmington Road, Mertztown, PA 19539

Plastics are finally gaining a position of eminence as premier protective barriers against hostile environments in the various process industries. Plastics are proven viable alternatives to such formidable corrosion barriers as rubber, glass, carbon and graphite, brick and ceramics, wood, metals and various alloys.

Prior to World War II, European process industries such as steel and metalworking, chemical and petrochemical, pulp and paper, etc. were more aggressive than their U.S. counterparts in the use of plastics in hostile environments. Plastics were recognized as possessing:

a) Excellent chemical resistance to many organic and inorganic acids, alkalies, salts and solvents.
b) High strength in relation to weight for producing piping and equipment of excellent structural integrity.
c) Sufficient functionality to permit formulating for a multitude of end properties, i.e., chemical and/or thermal resistance or both, certain physical properties, etc.
d) The properties to permit ease of molding, forming and fabricating, and

when compared to various metals and alloys were found to be very cost effective.

Polyethylene and polyvinyl chloride and epoxies and phenolics were the most popular thermoplastic and thermoset materials, respectively, of that period.

Today industry makes available a host of excellent chemical resistant plastics for a multitude of applications. This paper will discuss fabricated polyethylene, polyvinyl chloride and polypropylene for specific industrial applications.

0097–6156/83/0229–0141$06.00/0
© 1983 American Chemical Society

Design Criteria

The selection of the proper plastic for a given corrosion environment is not a simple process. It requires careful analysis of the properties of the plastic contemplated. Equally importantly, it requires careful analysis of what is to be protected (tanks, exhaust systems, piping, etc.) and the environment to which they are being subjected. The following "check points" will be of assistance in establishing the environment to which the plastic will be subjected and the subsequent plastic selected.

A) Identify all chemicals and their concentrations. pH, although helpful, does not provide sufficient information.
B) Maximum and minimum temperature exposure, including any anticipated thermal shock. Is the protection required indoors, outdoors, or both? Mother Nature introduces many challenges for plastics, i.e., sun, wind, rain, freeze-thaw, etc.
C) Physical factors such as positive and negative pressures, velocity, abrasion, impact, vibration, etc.
D) Is the application one of containing (process vessels, storage tanks, scrubbers, etc.), conveying (pipe, valves, fittings, etc.) or exhausting (hoods, ducts, stacks, etc.).
E) If the plastic is to be a fabricated insert for an existing structure, provide all pertinent dimensional data and the material of which it is constructed, i.e., steel, other metals, concrete, wood, etc.
F) Is process obsolescence imminent or is long term protection required? A significant consideration profoundly affecting economics.

Table I enumerates the various physical and thermal properties of these plastics. These factors are minimum requirements for the design engineer.

Tank design is the most challenging for the design engineer. Experience has proven to be the most positive as well as the most expensive for many fabricators of thermoplastics. Much of the design information from which formulas are derived is proprietary and jealously guarded. As can be appreciated, design factors can be voluminous when one considers the available plastics, creep properties of these plastics at various temperatures, all of which must be determined as a function of time.

The myriad of government mandates has required plastics for most industrial applications to be self-extinguishing and/or fire retardant. The plastics industry has been responsive to these mandates as well as those imposed by in-house standards of many

TABLE I
THERMAL AND PHYSICAL PROPERTIES

Property	Type I PVC	Linear Polyethylene	Homopolymer Polypropylene
Tensile Strength, psi.	6,800	4,000	5,200
Compressive Strength, psi.	10,000	2,500	6,000
Flexural Strength, psi.	14,000	2,000	6,000
Flexural Modulus, psi.	500,000	80,000	230,000
Impact Strength IZOD (ft. lb./in. notch)	0.88	2.5	4.0
Elongation % (at yield)	3	25	20
Coefficient of Expansion (in./in./^{o}F. x 10^{-5}	4.0	7.0	3.9 to 6.1
Thermal Conductivity (Btu./hr./sq. ft./in./^{o}F.)x 10^{-4}	1.33	3.17	1.3
Maximum Useable Temp. (^{o}F.)	150	180	200

companies. Polypropylene, for example, is available in three fire retardant grades. Briefly, they are described as follows:

V-0 Grade - Self-extinguishing with no dripping of sheet upon ignition.

V-1 Grade - Similar to V-0 Grade except in thicknesses of less than 1/8". It is highly unlikely that this grade would be used for fabricating self-supporting structures.

V-2 Grade - Self-extinguishing with some dripping permissible, however, the droplets must be self-extinguishing.

Flame retardant properties are not without sacrifice to chemical resistance, formulation flexibility and physical properties. For V-0 and V-2 grades, for example, physical properties are reduced approximately 20% and 10%, respectively.

Polyvinyl chloride, polyethylene and polypropylene have exhibited outstanding versatility to a multitude of hostile environments in the various process industries.

Table II provides a general indication of the chemical resistance that might be expected from these three plastics.

TABLE II
CHEMICAL RESISTANCE

Environment	PVC (TYPE I)	Polyethylene	Polypropylene
Acetic Acid, 50%	R	R	R
Aniline	N	R	R
Chromic Acid, 15%	R	R	R
Ethyl Acetate	N	C	C
Ethyl Alcohol	R	R	R
Hydrochloric Acid	R	R	R
Hydrofluoric Acid, 50%	R	R	R
Hydrogen Peroxide	R	R	R
Nitric Acid	R (Up to 68%)	R (Up to 40% Cold)	R (Up to 60% Cold)
Phosphoric Acid	R	R	R
Sodium Hydroxide	R	R	R
Sodium Hypochlorite	R	R	R
Sodium Sulfate	R	R	R
Sulfuric Acid	R (Up to 96% Cold)	R (Up to 50% Cold)	R (Up to 60% Cold)
Toluene	N	C	C

R = Recommended
N = Not Recommended
C = Conditional

It is to be noted that Table II does not indicate resistance at a specific temperature. Concentration and temperature are the most significant factors contributing to the success and longevity of any plastic fabrication.

Fabrication

Polyethylene, polyvinyl chloride and polypropylene are easily worked utilizing ordinary wood or metal working hand and power tools. Being thermoplastic, these materials can be easily thermoformed as well as cut, sawed, drilled and tapped. In the heat forming process, electric ovens are generally used which can maintain good thermal control. Of the three thermoplastics, polypropylene requires the more finite thermal control. Table III provides approximate time and temperature relationship required for thermoforming these three plastics.

When forming cylindrical stock for subsequent fabricating into ducts, stacks, tanks, etc. the sheets are removed from the oven and formed on mandrels. The sheet is held on the mandrels with a cotton canvas until cool. Polyvinyl chloride will cool faster than polyethylene and polypropylene, consequently, expediting the next step in the fabricating process.

TABLE III
TIME AND TEMPERATURE RELATIONSHIP

Plastic	Sheet Size	Temp., ^{o}F.	(^{o}C.)	Time, Mins.
	4' x 8' x 1/4"			
PVC		250	(120)	20
Polyethylene		270	(130)	40
Polypropylene		340	(170)	45
	4' x 8' x 1/2"			
PVC		250	(120)	30
Polyethylene		270	(130)	60
Polypropylene		340	(170)	80

The three plastics lend themselves well to heat welding. The three types of welding most commonly used are:

1. Butt Welding
2. Extrusion Welding
3. Hand Rod Welding

Butt welding, sometimes referred to as fusion welding, is used when fabricating large quantities of flat stock as well as large diameter or rectangular objects. Butt welding equipment is precise and highly automated. The equipment aligns, holds and welds the plastic at precise temperatures thus insuring welds of outstanding structural integrity. Butt welding generally eliminates the necessity to chamfer sheet edges prior to welding. Weld strength is at least 95% of the tensile strength of the virgin sheet.

Extrusion welding is done with guns that utilize resin pellets or reels of small diameter welding rod. Extrusion welding with pellets is used on essentially horizontal surfaces. Guns using rod are functional for horizontal, vertical and overhead welding. These guns generally utilize rod of 1/8" and 5/32" diameter. Extrusion welding cannot be used on stock thinner than 3/8" because the heat generated causes excessive sheet warpage. Sheets are generally chamfered on both sides to a total angle of 60° to accommodate welding the full depth of the sheet. Weld strength is at least 85% of the tensile strength of the virgin sheet.

Hand rod welding is generally done for small production runs and intricate shapes where the other methods of welding are not practical. Sheet edge chamfering is the same as for extrusion welding. Weld strength is at least 75% of the tensile strength of the virgin sheet.

Polyvinyl chloride welding can be done with hot air, however, polyethylene and polypropylene yield best results when welded with inert gas. Repeated laboratory studies have proven that nitrogen welding will provide the strengths alluded to above, notwithstanding of course the skills of the fabricators.

Applications

Polyvinyl chloride, polyethylene and polypropylene have established an enviable record of success in combatting corrosion in industry. These plastics possess outstanding versatility to hostile environments. The inherent differences in resistance to temperature and impact represent the most significant factors to the ultimate plastic selected for a given end use. Of the three plastics, polypropylene homopolymers offer the broadest range of chemical resistance combined with excellent physical properties. Outstanding chemical resistance with less physical and thermal demands would favor the use of polyvinyl chloride. Polyethylene exhibits excellent impact capabilities but due to its lower load bearing qualities, its use will generally be limited to small parts and equipment. Its low friction waxlike finish makes it attractive for use in anti-stick applications.

The following are typical applications as well as a brief description of the design of the various parts and equipment. The respective applications require uncompromising structural and physical capabilities with resistance to hostile environments characteristic of the chemical, steel and metalworking industries.

Steel and Metalworking Industries

A. Self-supporting Tube Pickling Tank (Figure 1)
Pickling Solution - 20% Sulfuric Acid at 180° F.
Material of Construction - Polypropylene Homopolymer
Tank Size - 32' Long x 5' Wide x 4' Deep
Shell Thickness - 1" Stock
All Corners - Molded to approximately 4" radius
Reinforcing - Exterior Only - bottom and sides with horizontal and vertical stiffeners of 3/4" and 1" thick polypropylene homopolymer stock.

B. Insert for Existing Carbon Steel Tube Pickling Tank (Figure 2)
Pickling Solution - 12% Nitric and 5% Hydrofluoric Acids at 130° F.
Material of Construction - Polypropylene Homopolymer
Tank Size - 50' Long x 3' Wide x 4' Deep
Shell Thickness - 1/2" Stock
All Corners - Rectangular to match existing tank
Interior Existing Tank - Protected with three coats vinyl coating. Sponge rubber approximately 1/2"

Figure 1. Self-supporting-tube pickling tank.

Figure 2. Insert for carbon-steel-tube pickling tank.

thick applied to interior to provide cushion and provision for expansion of the polypropylene insert.
Reinforcing - 1/2" thick vertical stiffeners on the interior of the polypropylene insert.

C. Covers for 80" Continuous Strip Pickling Line (Figure 3)

Covers prevent acid splash and provide confinement of fumes for subsequent exhausting.
Pickling Solution - 8% Hydrochloric Acid at 220° F.
Material of Construction - V-2 Fire Retardant Polypropylene Homopolymer
Tank Size - 372' Long x 10' Wide x 5' Deep
Cover Thickness - 3/4" Stock
All Corners - Rectangular
Reinforcing - Underside of covers using longitudinal and transverse stiffeners of 3/4" and 1" stock. Outside of covers are required to clear span.

D. Polyethylene Dipping Baskets (Figure 4)

Solutions - Plating Solutions at 140° F.
Material of Construction - Polyethylene, Linear
Size - 24" x 24" x 24"
Shell Thickness - 1/4"
Reinforcing - 1/2"

Chemical Process Industries

E. Chemical Hood, Cabinet and Sinks (Figure 5)

Service Conditions - Fume and Splash Control of hazardous and highly oxidizing chemicals
Material of Construction - Polyvinyl Chloride, Type I
Size - 60" Wide x 30" Deep x 60" High
Shell Thickness - 1/4" Stock
All Corners - Rectangular
Reinforcing - 1/2"

F. Dyestuff Filter Nutch (Figure 6)

Service Condition - Acid Pigments
Material of Construction - Polypropylene Homopolymer
Size - 60" Long x 36" Wide x 24" Deep
Shell Thickness - 3/4" Stock
All Corners - Molded to approximately 4" radius
Reinforcing - Exterior Only - Bottom and sides with horizontal and vertical stiffeners of 3/4" thick polypropylene homopolymer stock.

Summary

Figures 1 through 6 are pictures of the aforementioned

Figure 3. Covers for continuous-strip pickling line.

Figure 4. Polyethylene dipping baskets.

Figure 5. Chemical hoods, cabinets, and sinks.

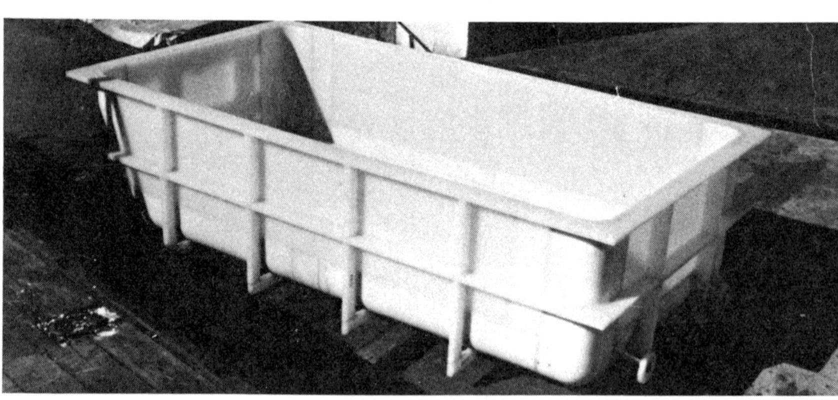

Figure 6. Dyestuff filter nutch.

installations which serve to illustrate the versatility of polyvinyl chloride, polyethylene and polypropylene.

On history, polyvinyl chloride and polyethylene have had over twenty-five years of successful applications in the United States with the successes expected to continue for years ahead. The success that these two materials enjoyed in Europe dates back to the pre-World War II era.

Polypropylene fabrication in the United States is in its infancy when compared to the other two thermoplastics. In the last five years, polypropylene has gained tremendous momentum for resistance to hostile environments, particularly where impact and structural integrity are paramount. One need only to recognize that the cover on a continuous strip pickling line, referred to in Figure 3 requires that a 300 lb. man may walk safely across this pickle line. This truly is confirmation of the tremendous confidence that is being placed in thermoplastics for not only resisting hostile environments but as well for their ability to perform structurally. The process industries (chemical, steel and metalworking, to mention a few) will continue to depend heavily on the use of these three custom fabricated thermoplastics to assist them in combatting corrosion.

RECEIVED March 3, 1983

10

Water Permeation Through Elastomers

PATRICK E. CASSIDY

Texas Research Institute, Inc., Austin, TX 78746, and Southwest Texas State University, San Marcos, TX 78666

Long term monitoring of deionized (DI) and salt water through Neoprenes G and WRT reveal an increasing permeation rate with age for DI water. Several explanations for this behavior are offered, with supporting analytical data. The effects on permeation of elevated temperatures and/or static and dynamic mechanical stresses are examined. Temperatures of 60-70°C cause dramatic increases in permeation rate for both DI and salt water, with relative values depending on the rubber composition. Small static stresses were ineffectual. The dynamic stress applied--high frequency acoustic agitation--increased permeability through heating.

Plastics and elastomers are generally thought to be resistant to or unaffected by mild environments such as water immersion. Commonly, an organic coating polymer or seal is accepted as providing a water-tight barrier even when the device is immersed in water. But water will permeate the barrier, if only slowly. The phenomenon of water permeation is reviewed in (1,2).

There are three factors that can adversely affect a material employed as a barrier: time, temperature, and mechanical stress. While the last two are usually considered when designing a piece of hardware, the first is not always taken into account. Part of the reason for disregarding time is the lack of reliable aging information.

In order to evaluate a material's lifetime expectancy in a short laboratory test the temperature is increased and a temperature-time relationship is developed for some property of the material. Of course, there is a limit to how high the temperature can be (e.g., 80-90°C for Neoprenes) without inducing extraneous and unrealistic chemical reactions. More difficult sometimes is the decision criterion: when the decay of the

0097-6156/83/0229-0153$06.00/0
© 1983 American Chemical Society

material property (e.g., tensile strength) is such that the material is considered failed for a specific application.

This paper considers Neoprene elastomers and their use as barriers to water penetration. Long time effects on permeability and composition will be discussed as will the effects on permeability of temperature and high-frequency mechanical (acoustic) agitation.

Discussion

Effects of Long-Term Water Exposure. The usual procedure for evaluating the permeability of an elastomer is to monitor water transmission rate through the rubber until a steady state is attained (ASTM D 1653), usually a period of a few days. This measured permeation rate is then used to project total water transmission over the expected lifetime of a device, which may be several years (3). This is valid, of course, only if no critical changes (e.g., degradation, leaching of components, hardening, etc.) take place in the elastomer as it ages. However, several aging reactions are known. The most severe is perhaps the extraction of components of the rubber into the liquid permeant. Another is the continued swelling of the elastomer due to increased dissolution of the permeant into the membrane. In any case, long-time extrapolations of permeability rates determined over short durations may not be valid (4).

Indeed, long-term measurements of deionized (DI) water permeation through samples of Neoprene G (316 days at 60°C) and Neoprene WRT (75 days at 25°C) gave nonlinear permeation curves. Figure 1 illustrates this for Neoprene G. Shown are plots of weight loss versus time in permeation cells filled with DI water and with 3.5% salt water. The latter yielded a straight-line plot, i.e., constant permeation rate, which verifies the technique and is explained by the fact that salt water causes only a very small change in the composition of the rubber. The rate of DI water permeation was 139% greater at the end of the test period than at the beginning. A Neoprene WRT sample showed a gain to 450% of its original value, and this under even milder conditions than the Neoprene G test (5). The Neoprene G results are tabulated in Table I.

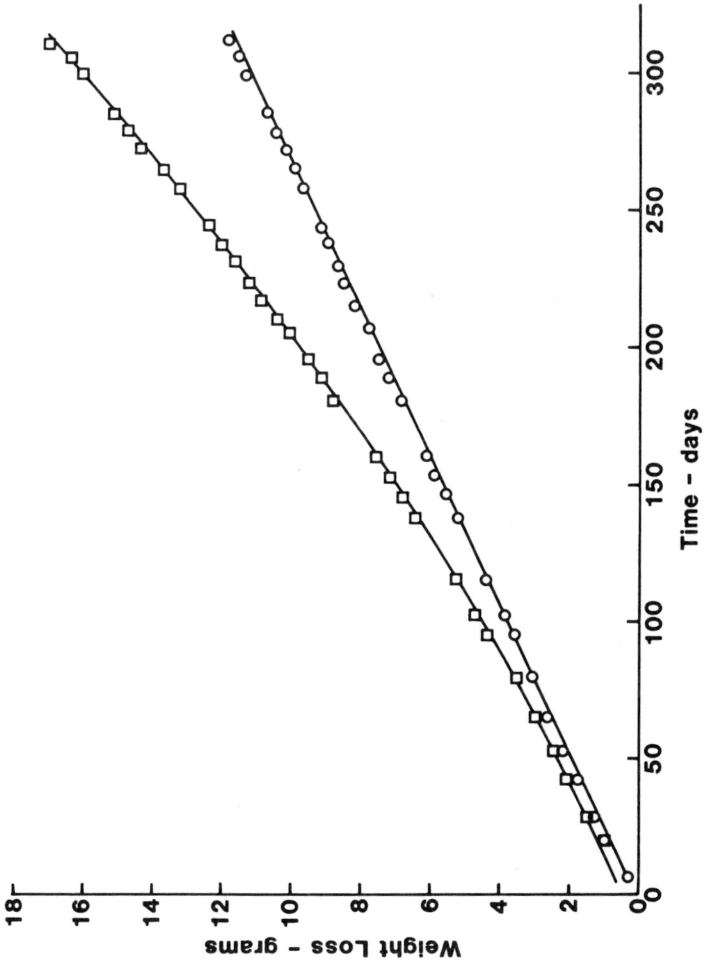

Figure 1. Water permeation through Neoprene G at 60°C. Key to permeant: □, *deionized water; and* ○, *3.5% NaCl solution.*

Table I. Long-Term Permeation Results

	Total Time (days)	Total Water Permeated (g)	Conductivity of Face Wash# ($ohm^{-1}cm^{-1}$)	Permeation Rate† ($mg/cm^2/day$)
DI	316	16.53	8.6×10^{-6}	1.69/2.35
SALT	316	11.51	7.6×10^{-6}	1.43

Pure DI water is 5×10^{-6} $ohm^{-1}cm^{-1}$
† For DI water, initial and final rates are given

There are several possible explanations for this nonequilibrium behavior. The composition of the Neoprene may be changed by the leaching of components into the liquid permeant, thereby opening pores to increased capillary flow of water (vis-a-vis diffusion). Changes in composition of the elastomer and the residual permeant are discussed in later paragraphs.

Another possible explanation may be increasing solubility of DI water in these Neoprene formulations, due, of course, to the additives. Other studies carried out at 5, 25 and 50°C over 24-40 day terms have shown that fresh water continues to be absorbed by Neoprene WRT without reaching equilibrium (vide infra). Because permeability is dependent on solubility, a continuing increase in the latter will affect the former similarly.

A third possible explanation is that absorption of water in elastomers which contain water soluble fillers has been shown to create pockets of solution which continue to grow until the osmotic pressure is balanced by the containment pressure of the rubber. This introduces a very complex picture for water permeation: there are pockets of permeant, which cause mechanical stresses in the rubber matrix, creating thereby channels for water diffusion.

In the Neoprene G tests summarized in Figure 1 and Table I the outer (dry) faces of the permeated rubber samples were washed with DI water to determine if soluble salts were present. The conductivity values listed in Table I show only a very slight difference, which is probably of little import. The internal water permeant was, however, changed by the long term aging. The water was pale yellow. Analysis by emission spectroscopy showed notable changes in the concentrations of 4 of the 16 metals which were monitored. These results are listed in Table II.

Table II. Metal Concentrations (%) in Residual Permeants

Metal	Concentration (%) Deionized Water	Salt Water
Al	7×10^{-5}	5×10^{-2}
Mg	0.18	0.05
Na	0.01	1.9*
Fe	7×10^{-4}	3×10^{-3}

*Corresponds to 4.8% NaCl solution.

Aluminum appeared in the residual salt water permeant at a concentration 700X that in the DI water. Inspection of the permeation cells revealed much more internal corrosion associated with the salt water permeant than with the fresh water; hence, the source of aluminum. Sodium, of course, originated from the Neoprene, and salt water removed much less than did fresh water. The high concentration of sodium ion in the residual salt water permeant (4.8% vis-a-vis the initial 3.5%) resulted from the loss of water by permeation while retaining salt. The source for magnesium and iron was also the Neoprene.

The permeated rubber was also analyzed for metals by emission spectroscopy. Of the 16 metals present only sodium showed any appreciable change in concentration. Salt water reduced the sodium content of the rubber by 17% (from 0.12% to 0.10%) whereas DI water reduced it by 92% (to 0.01%).

Two samples of Neoprene G were sliced into 40 μm sections with a cryogenic microtome to examine ion migration. The samples were from an unused piece of material and one which had been subjected to salt water permeation at 60°C for 316 days.

Several slices of each sample were analyzed by ESCA (also called XPS) for cation content. Figure 2 shows the relative abundance of several elements versus depth in the rubber. (Note that the logarithmic scale deemphasizes the differences along the y axis). A possible movement of species in the direction of flow (right to left) is apparent in this graph. Or, species may be removed by leaching by the permeant, which would result in low values on the wet side. The latter argument is weakened by the fact that at slice No. 16, 87% of the way through the rubber, the abundance of most of the species exceeds their original value.

Effects of Elevated Temperature

Most rate processes increase with increasing temperature, for example, the solubilities of solids in liquids. Of course, there are exceptions, e.g., the solubility of water in polyimides (6); and with plastics and elastomers there are upper temperature

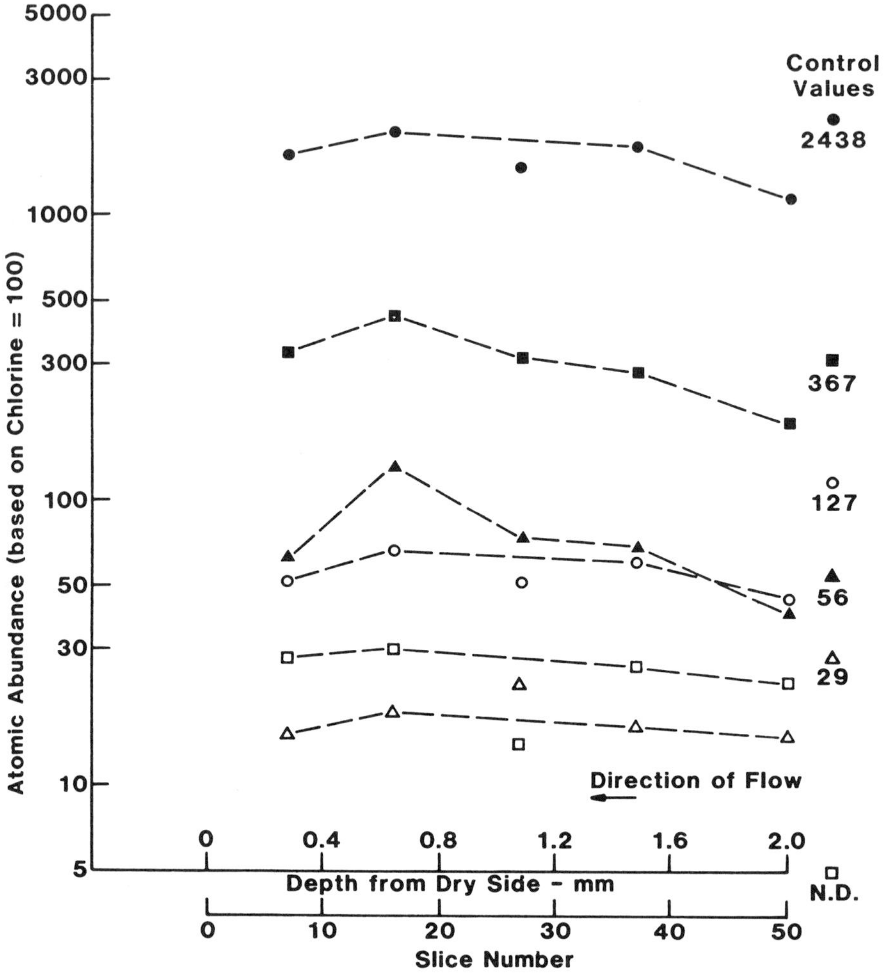

Figure 2. Element abundance vs. depth in sample. Key: ●, carbon; ■, oxygen; ▲, aluminum; ○, silicon; □, sodium; and △, sulfur.

limits above which the processes themselves may be altered. Solubilities and permeabilities have been evaluated for Neoprenes (3,7), polyisoprenes and natural rubbers (8) for temperatures to 60°C.

Figure 3 illustrates the change of solubility of water in WRT Neoprene with temperature. The solubilities of both deionized and 3.5% salt water were measured at 5, 25 and 50°C. The figure shows that DI water is more soluble than salt water in WRT Neoprene and that increasing temperature increases solubility. The solubility of DI water at 50°C does not approach equilibrium even after 25 days. From these data a total weight gain of 300% was predicted.

Assuming that diffusivity behaves similarly, permeability would be expected to increase with temperature.

Several elastomers were evaluated for permeability to salt and deionized water at several temperatures. A WRT Neoprene (Burke Rubber Co. Type 5109) was tested at 23, 40 and 70°C using both salt water (3.5%) and DI water as permeants. The results for salt water are listed in Table III and for DI water in Table IV. Salt water permeability increased 24 times with a 47°C temperature rise while the DI water increase was 33 times. The difference may be due to the much greater solubility of DI water relative to salt water.

Table III. Permeability of Burke 5109 Neoprene to Salt Water

Temp. (°C)	Permeant	Permeability (g-cm/cm^2-day)x10^6	Permeation Constant (ng-cm/cm^2-hr-torr)
23	3.5 wt-%	33.3	66.4
40	NaCl	55.6	42.3
60	Solution	190	53.7
70		792	142.7

Table IV. Permeability of Burke 5109 Neoprene to DI Water

Temp. (°C)	Permeant	Permeability (g-cm/cm^2-day)x10^6	Permeation Constant (ng-cm/cm^2-hr-torr)
23	DI	13.3	26.3
40	Water	50.0	37.6
60		255	71.2
70		433	77.2

Energies of activation determined from an Arrhenius plot of these data yielded values of 12.8 kcal/mole for the NaCl water permeant samples and 15.3 kcal/mole for the DI water permeant samples.

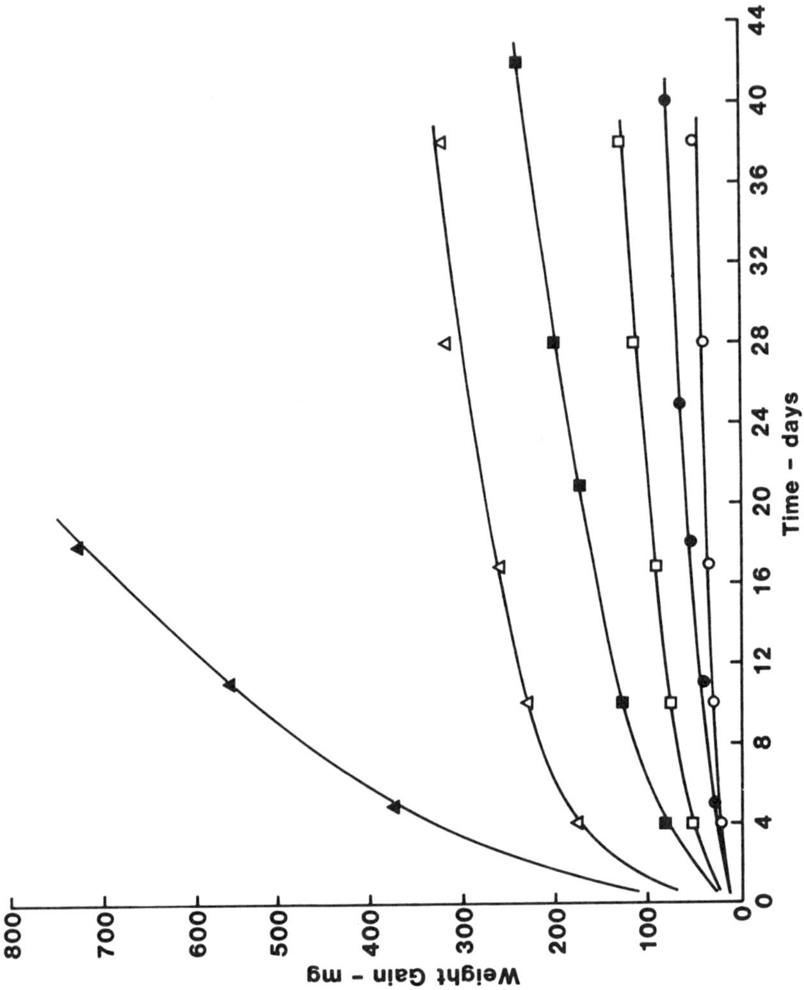

Figure 3. Evolution of the solubilities of deionized water (solid symbols) and 3.5% NaCl solution (open symbols) in WRT Neoprene. Key to temperatures: ○●, *5°C;* □■, *25°C; and* △▲, *50°C.*

Another Neoprene (Type CR, Utex Industries) showed increased permeability with temperature, but in this case salt water was more permeating than DI water (Figure 4). Interestingly then, the relative rates of salt and DI water permeation of Neoprene can be inverted depending on rubber composition.

Note in Figure 4 that as temperature increases the difference in rates between DI and salt water decreases.

Analogous results for EPDM and SBR are presented in Figures 5 and 6. Here again, salt water is more permeating than deionized water.

Specific data from Figures 4, 5 and 6, including activation energies, are summarized in Table V.

Figures 7 and 8 display the preceding results in a different format: permeation rate is plotted against temperature for DI and salt water, respectively. In each case Neoprene is by far the most permeable and EPDM the least. An exponential increase of rate with temperature is also obvious.

Effect of Mechanical Stress on Permeability

There are several underwater uses of elastomers in which they are subjected to mechanical stresses. This can be static wherein the rubber is stretched and held at a certain elongation while in service. Or it can be dynamic wherein a seal, for example, is subjected to mechanical agitation, vibration, etc.

It is reasonable to assume that a mechanical stress would supply energy to the elastomer thereby increasing the rates of any processes. For example, in aging, plastics or elastomers which are highly elongated will decompose more rapidly than unstressed controls. There is, however, no published information on the effects of ultrasonic radiation on water permeation through elastomers.

Experiments were run to evaluate the effects of low level static (stretch) and dynamic (ultrasonic irradiation) stress on water permeation of a WRT Neoprene. At the same time, the contribution of a partial silica filler in the place of black was determined. The materials were identical zinc oxide-magnesium oxide-cured formulations except that one contained 30 phr black and the other contained 15 phr black and 15 phr silica. These were molded into cups with 0.030 in. thick faces and 0.070 in. thick side walls which were then fitted over 3.5% salt water-filled glass cups and sealed with hose clamps.

To obtain a static stress (stretched membrane), the face area of the glass cups was made 10% larger than the face area of the rubber cups.

The dynamically stressed samples were not stretched, i.e., the exterior dimensions of the glass cup matched the interior dimensions of the rubber cup. An ultrasonic transducer was

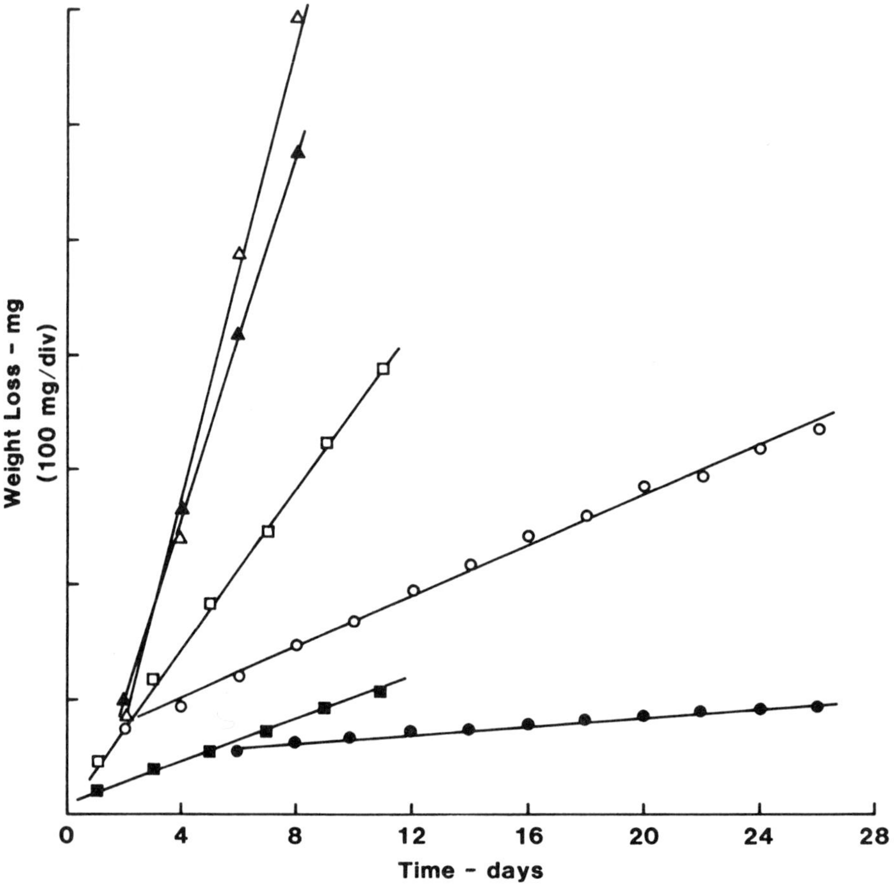

Figure 4. Effect of temperature on permeation of deionized water (solid symbols) and 3.5% NaCl solution (open symbols) through CR Neoprene. Key to temperature: ○●, 23°C; □■, 40°C; and △▲, 60°C.

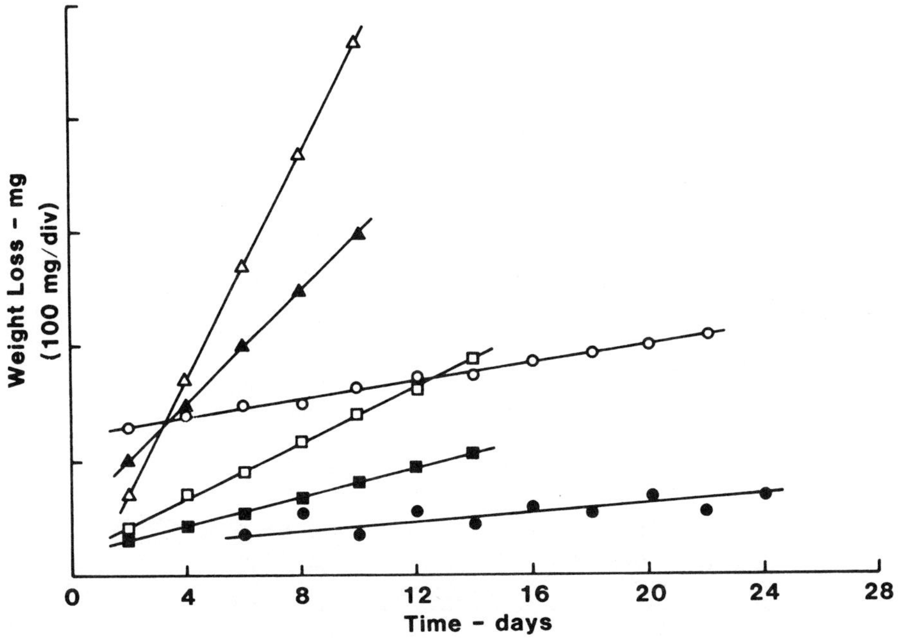

Figure 5. Effect of temperature on permeation of deionized water (solid symbols) and 3.5% NaCl solution (open symbols) through EPDM. Key to temperature: ○●, 23°C; □■, 40°C; and △▲, 60°C.

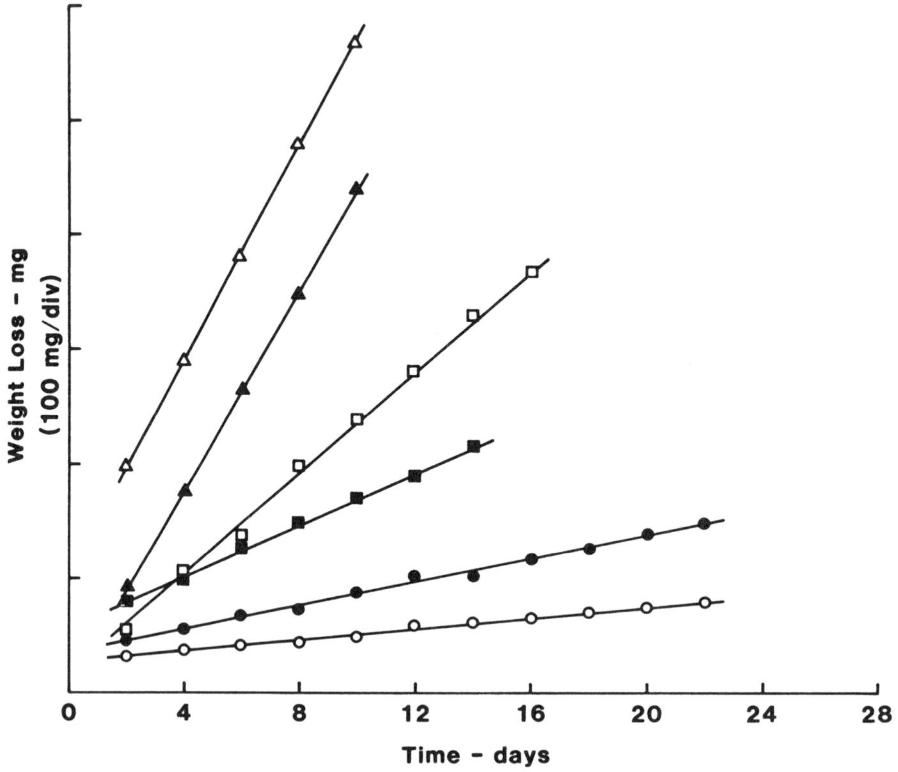

Figure 6. Effect of temperature on permeation of deionized water (solid symbols) and 3.5% NaCl solution (open symbols) through SBR. Key to temperature: ○●, 23°C; □■, 40°C; and △▲, 60°C.

Table V. Permeability Data on Various Elastomers Exposed to Distilled Water and 3.5% Aqueous NaCl at 23°, 40° and 60°C

Material	Permeability Coefficient $(g\text{-}cm/cm^2\text{-}sec\text{-}cmHg)\times10^{11}$ 23°		40°C		60°C		E_p (kcal/mole)	
	Salt Water	Distilled Water	Salt Water	Distilled Water	Salt Water	Distilled Water	Salt Water	Distilled Water
CR	54.94	8.97	64.79	17.82	66.86	54.28	11.8	19.9
SBR	10.65	6.17	10.90	10.49	15.53	15.14	12.0	15.1
EPDM	3.93	1.68	4.40	2.31	6.60	4.50	12.7	17.1

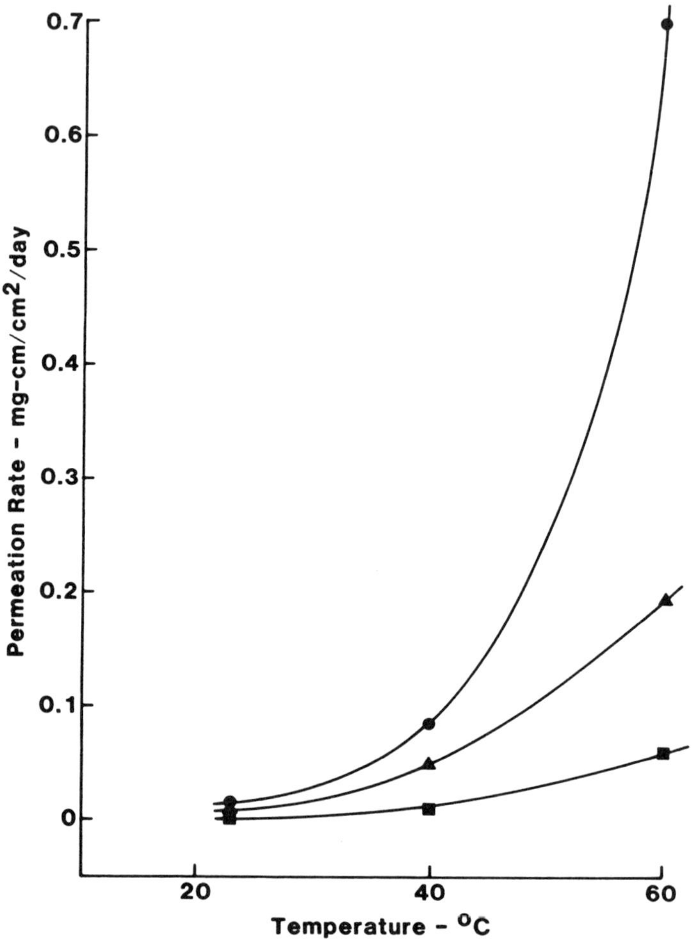

Figure 7. Temperature dependence of permeation rate for deionized water. Key to material: ●, CR Neoprene; ■, EPDM; and ▲, SBR.

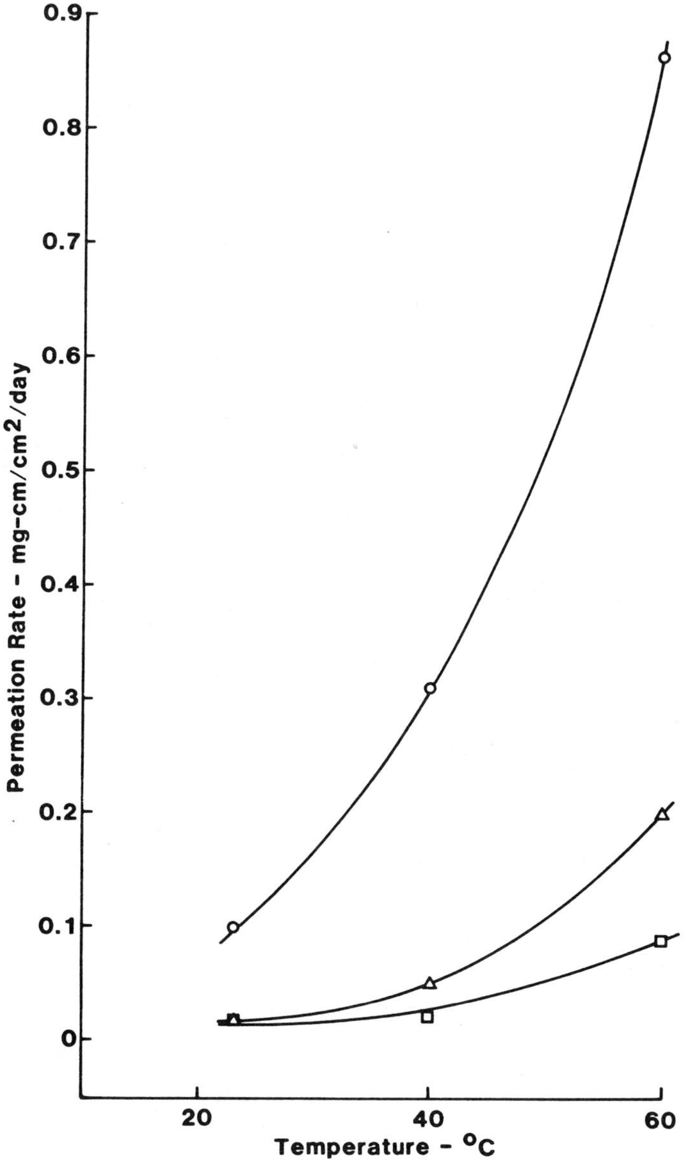

Figure 8. Temperature dependence of permeation rate for salt water. Key to material: ○, *CR Neoprene;* □, *EPDM; and* △, *SBR.*

coupled to the bottom of the glass cup through a thin slab of silicone rubber. Two transducers were used, one resonant at 52 kHz and one at 99 kHz. Both were driven at their resonant frequency with a power input to the cup of 0.08 W/cm^2. Permeation was monitored over a period of 2-4 days. For measurement, the transducer was removed from the cup assembly, which was then cooled to room temperature and weighed to the nearest 0.01 mg. A permeability coefficient was calculated from the slope of the weight loss versus time plot.

Table VI gives salt water permeability coefficients obtained at 60°C for the control samples and the stretched samples, and at an unknown, but higher temperature (probably about 75°C) for those samples which were ultrasonically stressed (ensonified and thereby heated) during permeation. It is apparent that the presence of silica caused no difference in permeability, nor did the static stress of 10% stretch (when corrected for the 10% decrease in thickness which occurred). However, a very striking result was that ensonification yielded a 370% increase in permeability coefficient, which could have been due to the excess heating.

Table VI. Salt Water Permeation in Stressed Neoprene

Control Samples (60°C)	Permeability Coefficient ($mg\text{-}cm/cm^2\text{-}day\text{-}mmHg$) x 10^3	% of Original
NXA-1883 Silica Filled	14.62	--
NXA-1938 Non-silica Filled	14.15	--
Statically Stressed Samples (60°C)		
NXA-1883 Silica Filled	14.15	97
NXA-1938 Non-silica Filled	14.15	100
Dynamically Stressed Samples Ensonified at 99 kHz; Temperature about 75°C		
NXA-1883 Silica Filled	52.6	360
NXA-1938 Non-silica Filled	52.6	372
After Dynamic Stress (60°C)		
NXA-1883 Silica Filled	12.18	83
NXA-1938 Non-silica Filled	11.92	84

Subsequent tests performed using a thermistor attached to the permeation cell revealed that although the cell was in a desiccator in a 60°C oven, heat generated by the sound source raised the temperature of the cell to about 75°C due to the slow rate of heat dissipation through the desiccator. To determine whether this increased temperature sould account for some of the increased permeability, similar tests were run using the transducer resonant at 52 kHz; this allowed assessment of a frequency effect, if any, as well. These tests again showed an increase in cell temperature to 70–72°C, even though the oven chamber was at 60°C. Evidently, the energy generated by the transducer cannot be rapidly dissipated through the desiccator.

Therefore, permeation rates were measured for the ensonified material in a 60°C oven (70–72°C cell temperature) and for a control at 72°C. The results are shown in Table VII.

It appears that the increased permeation during ensonification is due to heating of the sample by the sound source. The differences seen in Table VII can be accounted for by the 2°C variation in temperature of the cell during ensonification. Whether there is a frequency (or wavelength) dependence beyond the range of frequencies tested remains to be determined.

Also evident in these data is a decrease (nom. 16%) in permeability relative to the control values after ensonification. Apparently the latter caused some change in the Neoprene.

Conclusions

Time, temperature, and mechanical stress must all be considered when an elastomer or plastic is to be used as a barrier to water or some other permeant. Long time exposures can cause changes in composition of a barrier which lead to increased solubility of the permeant and to higher permeation rates.

Temperature effects are important, not only in high temperature applications, but also during accelerated aging of a material for laboratory study and subsequent evaluation of in-service life-time. Solubility of water in Neoprene increases greatly with temperature as does permeation rate. The relative permeation rates of 3.5% saltwater and deionized water were shown to depend on elastomer composition.

Although small (10%) static stresses (stretching) have no effect on the permeability of Neoprene, ultrasonic agitation increases salt water permeation rates severalfold due to heating by ensonification.

TABLE VII. Permeation Through Neoprene
(3.5% NaCl Solution; 52 kHz Ensonification)

Control (72°C	(mg/hr)	Permeability (mg cm/cm^{2}day cmHg x 10^3)	% of Control
NXA-1883 Silica Filled	3.94	23.2	
NXA-1938 Non-Silica Filled	3.17	18.5	
Ensonified (70-72°C)			
NXA-1883 Silica Filled	4.08	23.9	+3.4%*
NXA-1938 Non-Silica Filled	2.92	17.1	-8.7%*

*due probably to temperature variation (2°C) during ensonification

Acknowledgments

Portions of this work were funded by the Naval Research Laboratory/Underwater Sound Reference Detachment, Orlando, Florida and by the Robert A. Welch Foundation, Houston, Texas.

Literature Cited

1. Barrie, J. A. in "Diffusion in Polymers," Crank, J.; Park, G. S., Eds.; Academic: New York, 1968; p. 259.
2. Cassidy, P. E.; Aminabhavi, T. M.; Thompson, C. M. in "Rubber Chemistry and Technology," in press.
3. Cassidy, P. E.; Perry, R. B. SAMPE Quarterly 1975, 6, 34.
4. Cassidy, P. E.; Rolls, G. C. J. Appl. Polym. Sci. 1982, 27, 2743.
5. Thompson, C. M., personal communication.
6. Thompson, C. M. J. Appl. Polym. Sci. 1981, 26, 373.
7. Cassidy, P. E.; Rolls, G. C. Texas J. Sci. 1972, 24, 325.
8. Barrie, J. A.; Machin, D.; Nunn, A. Polymer 1975, 16, 811.

RECEIVED April 7, 1983

11

Polyurethane Aging in Water and Methanol Environments

DOUGLAS L. FAULKNER, MICHAEL G. WYZGOSKI, and MARK E. MYERS, JR.

General Motors Research Laboratories, Warren, MI 48090–9055

The tensile properties of polyester-based thermoplastic polyurethanes were studied as a function of the time of exposure of the plastics to water, methanol, methanol-water, methanol-isooctane, and methanol-water-isooctane. The resulting decrease in the tensile properties of the plastics was attributed to reaction of the plastics with water and methanol. As indicated by the decrease in properties, reaction with methanol is initially faster, but the reaction rate with water increases with time -- presumably because of the autocatalytic nature of the reaction. Nuclear magnetic resonance spectroscopy indicated that the reaction mechanisms with methanol and water were transesterification and hydrolysis, respectively. Gel permeation chromatography revealed that both reactions resulted in a marked reduction in the molecular weight distribution of the plastics, which correlated well with losses in tensile properties.

Mechanisms for polyurethane degradation include hydrolysis and chemical exchange reactions. Hydrolysis involves a reaction of the polymer with water which results in chain cleavage. Chemical exchange is a general term which includes such reactions as glycolysis, aminolysis, transesterification and the like. These reactions, like hydrolysis, result in molecular weight reduction through random chain scission. Ultimately all of these reactions will cause an attendant loss of tensile properties.

This paper presents results from aging polyester based polyurethanes in methanol and water environments and in other mixtures of these media. The influence of aging on mechanical properties was assessed by measuring tensile strengths and elongations at

0097–6156/83/0229–0173$06.00/0
© 1983 American Chemical Society

various exposure times. Chemical analyses were used to independently monitor the extent of degradation and to determine the aging mechanisms.

Experimental

Materials. Injection molded Texin 591AR thermoplastic polyurethane, supplied by Mobay Chemical Company, was one of the two polyurethanes used in this investigation. (Hereafter this resin is referred to as polymer "A".) This resin is a polyester based polyurethane and is considered a medium grade in terms of hardness or flexibility. Nuclear magnetic resonance (NMR) characterization revealed that for polymer A the molar ratio of polyester to diphenylmethane-4,4'-diisocyanate (MDI) is 2.07, and the ester is poly(butylene adipate). The material was obtained in pellet form to prepare samples for aging experiments.

Pellethane CPR 2102 polyurethane, obtained from the Upjohn Company, was the other polyurethane studied. (Hereafter this material is referred to as polymer "B".) For polymer B the molar ratio of polyester to MDI is 2.95, and the ester is poly(caprolactone). This material is reported to have initial properties similar to polymer A when both materials are post cured at 110°C for 16 hours. The polymer B samples were previously post cured. In addition, polymer B is reported to provide superior resistance to hydrolytic attack compared to other polyester based polyurethanes.

Distilled water, isooctane, and methanol were used as aging media. Isooctane and methanol were reagent grade and were used as received without further purification. Isooctane was selected as a relatively inert fluid to study the effect of methanol concentration.

Tensile Sample Preparation. Polymer A tensile bars were injection molded according to the supplier's recommendations. Dimensions were 3.2 by 3.8 mm with a length of 62.0 mm. Polymer B was supplied in 1.7 mm thick sheets from which 3.2 mm wide tensile bars were cut.

Aging of Tensile Bars. Tensile bars were aged in test tubes containing the aging media. The test tubes were heated in an aluminum block bath and were equipped with water cooled condensers to minimize liquid loss through evaporation. All samples were aged at 80°C except samples aged in methanol, where the aging temperature was 60°C to prevent the boiling of methanol (b.p.=64.5°C). All aging liquids were miscible when heated except for a mixture of water, methanol, and isooctane, which formed a two-phase mixture. In this case, the samples were suspended in the isooctane-rich phase. After aging, samples were dried in a vacuum oven at 80°C and reequilibrated at room temperature under

50% relative humidity before tensile testing. At the termination of these experiments, the various solvent mixtures used to age the samples were evaporated to dryness, and the resulting residues were saved for NMR analysis.

Tensile strengths and elongations were measured at room temperature using tensile specimens with a gauge length of 15.9 mm. An Instron was employed with a crosshead speed of 508 mm/min.

Chemical Analysis

Nuclear Magnetic Resonance (NMR). Samples were dissolved in dimethyl sulfoxide (d_6) at a concentration of about 100 mg/cc. The polyurethanes were completely soluble and did not require heating; however, the time required to dissolve them appeared to be strongly dependent upon the extent of degradation of the sample, with highly degraded samples dissolving more quickly. In most instances, samples were put on a shaking table overnight to achieve complete dissolution. The solutions were transferred to 5-mm diameter NMR tubes and 4 drops of tetramethylsilane (TMS) were added to provide field/frequency locking and to provide a reference line in the spectrum. Deuterochloroform was used as a solvent for many of the evaporation residues described in the previous section. Proton NMR spectra and integrals were obtained on a Varian HA-100D NMR spectrometer.

Gel Permeation Chromatography (GPC). Gel permeation chromatograms were obtained on a Waters 150-C liquid chromatograph operated in the GPC mode. Dimethyl sulfoxide was used as the solvent, and sample concentrations were about 0.25%. This instrument contains a microprocessor which computes the number and weight average molecular weights using a calibration curve generated from GPC chromatograms of polystyrene standards in tetrahydrofuran. The values of molecular weight obtained in this manner will not be true values; however, they can be used for comparisons among samples.

Liquid Sorption. Liquid sorption experiments were conducted in methanol-water and methanol-isooctane solutions at 60°C to assess the swelling characteristics of polyurethane in these media. Samples were approximately 3.0 mm thick and weighed about 0.9 gram. Gravimetric measurements were performed on a Sartorius analytical balance to the nearest 0.1 milligram. The samples were weighed dry and placed in the liquid for 24 hours. Samples were removed, blotted with tissue to eliminate adhering surface liquid, and rapidly placed in tared weighing bottles. Immediately following this procedure, the bottle and sample were reweighed. The difference from the original dry weight is the weight gain.

Results

Aging of Tensile Bars. Polyurethane tensile samples were aged in methanol, water, an equal volume mixture of the two, and other mixtures containing isooctane. While only graphical results of polymer A tensile properties are presented in this section, similar results were obtained with polymer B. Tensile properties of both polymers A and B are listed in Table I.

A plot depicting the effect of aging in water on the tensile strength of polymer A is shown in Fig. 1. The plot is characterized by an initial slow decrease in strength with time which accelerates after 10 days. This increase in the rate of tensile strength loss is consistent with the autocatalytic nature of the hydrolysis reaction. The reaction generates an acid which in turn catalyzes further hydrolysis of the polyurethane (1).

As shown in Fig. 2, tensile strength of polymer A decreases more rapidly initially in methanol than in water even though aging in methanol is conducted at 60°C (20°C lower than in water). The rate of property loss is observed not to increase with time. This is expected since the degradation reaction in methanol (transesterification) is not autocatalytic. The reduction of tensile elongation in the two media shows a similar behavior.

Tensile properties of the polyurethanes aged in solvent mixtures containing various amounts of methanol were also evaluated. The relative effects of the composition of the mixtures on the tensile strength of polymer A are shown in Fig. 3. The data appear to be systematic when expressed in terms of mole fraction of methanol (indicated in Fig. 3 in parenthesis). A two-phase mixture resulted upon the addition, on a volume basis, of a 25% aliquot of an equal volume methanol-water solution to isooctane (in an attempt to form a 75-12.5-12.5 isooctane-methanol-water solution). No volume change was observed in the isooctane layer upon the addition of the methanol-water solution. Given the polarity dissimilarities of water and isooctane, it is reasonable to expect that the methanol partitioning would favor much higher methanol concentrations in water than in isooctane. Since the samples were exposed only to the upper isooctane layer, which presumably contained little methanol, less degradation occurred here than in other mixtures containing higher methanol contents.

Chemical Analysis

NMR. The NMR spectrum of an unexposed sample of polymer A, shown in Fig. 4, indicates that the polymer contains MDI, ethylene glycol, and poly(butylene adipate). The assignments are as shown (2), and the various sets of bands are given letter designations. Chemical shifts are indicated as parts per million (ppm) from TMS.

Table I. Tensile Properties of Polyurethanes

Aging Medium*	Aging Time (Days)	Tensile Strength (MPa)		% Elongation	
		Polymer A	Polymer B	Polymer A	Polymer B
Control	-	25.85	51.71	1060	1020
Methanol	10	5.23	43.64	219	1220
"	20	2.36	13.03	32	448
"	30	0.59	11.38	16	368
Isooctane-methanol (75-25)	10	6.70	35.85	347	1070
"	20	2.76	15.10	107	592
"	30	0.93	8.89	32	261
Methanol-water (50-50)	10	14.89	39.99	955	1050
"	20	10.27	25.30	640	949
"	30	7.86	16.89	331	731
Isooctane-methanol-water (75-12.5-12.5)	10	22.61	47.78	1060	1030
"	20	18.82	46.75	907	1070
"	30	18.27	39.16	850	1030
Water	10	23.92	43.78	1090	1010
"	20	15.65	23.65	928	869
"	30	7.31	8.55	299	251

* Aging temperatures were 80°C except for methanol where the temperature was 60°C.

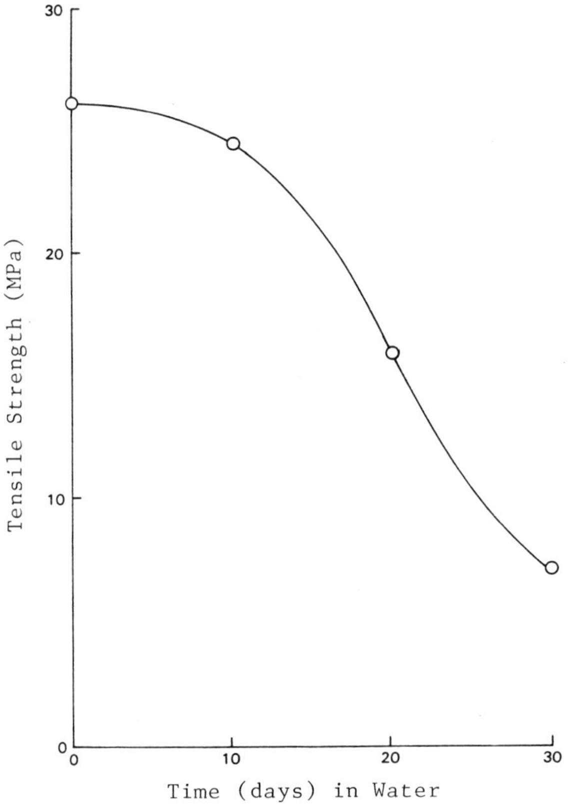

Figure 1. Kinetics of tensile strength loss of polymer A aged in water at 80°C.

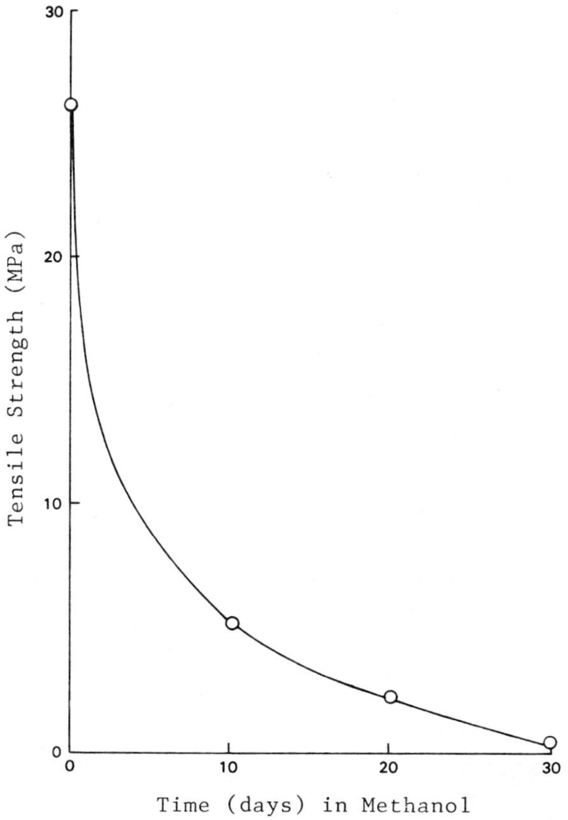

Figure 2. Kinetics of tensile strength loss of polymer A aged in methanol at 60°C.

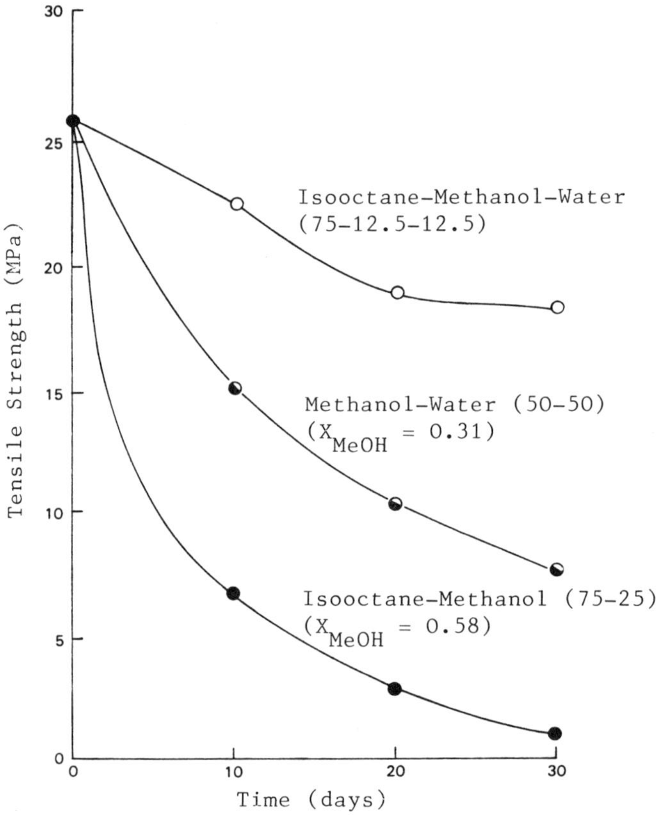

Figure 3. Kinetics of tensile strength loss of polymer A aged in methanol solvent mixtures at 80°C. X_{MeOH} is the mole fraction of methanol.

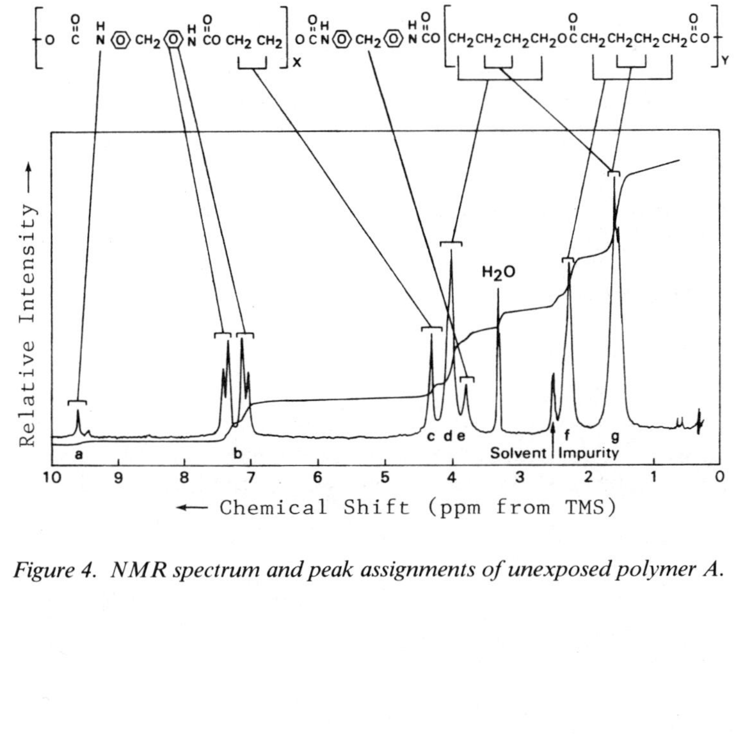

Figure 4. NMR spectrum and peak assignments of unexposed polymer A.

In methanol or water environments the polymer is expected to cleave at the ester linkage as shown below:

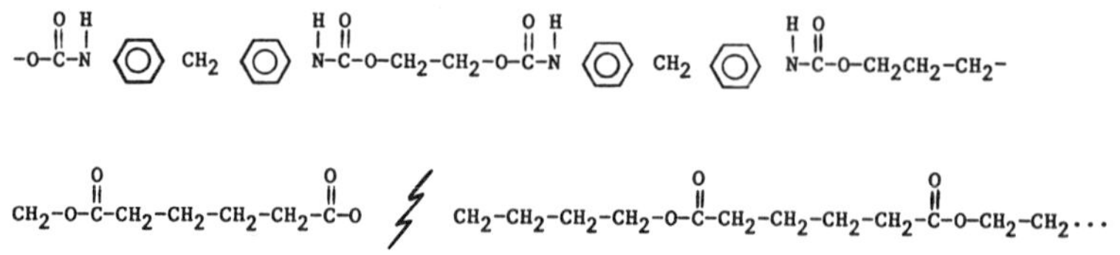

If methanol is present, the methyl ester should form at the cleavage points (transesterification) and act as a "tracer" in the NMR spectrum. If water alone causes the cleavage (hydrolysis), no new bands will appear in the spectrum because the acid OH formed is difficult to detect by proton NMR. In the present study the expected methyl esters were sometimes observed in the polymer after aging in methanol environments. Figure 5 shows the methyl ester peak observed in an NMR spectrum of polymer A aged 30 days in methanol.

It is possible to obtain quantitative data from the NMR spectrum integral. Referring to Fig. 4, if the integral of peaks designated b is divided by 8 (the number of protons in the MDI contributing to this resonance), we have a number for the relative amount of MDI present, that is,

$$\text{relative moles MDI} = b/8$$

In a similar manner, we can obtain a number for the relative amount of poly(butylene adipate) present by dividing the integral of peak g by 8 (the number of protons in the poly(butylene adipate) contributing to this resonance), that is,

$$\text{relative moles polyester} = g/8$$

Thus, we can obtain a value for the ratio $\frac{\text{moles polyester}}{\text{moles MDI}}$:

$$\frac{\text{moles polyester}}{\text{moles MDI}} = \frac{g/8}{b/8} = \frac{g}{b}$$

Figure 6 shows that the molar ratio of polyester to MDI decreases with aging time in methanol. In methanol, the degraded polyester is sufficiently soluble in the medium and is subsequently extracted from the polyurethane substrate (as will be shown later). The polyurethane is also highly swollen by methanol, thereby enhancing the rate of diffusion of polyester chain segments out of the polyurethane matrix. In contrast, when the material is exposed to water (see Fig. 7) the molar ratio is practically unchanged even though tensile properties have

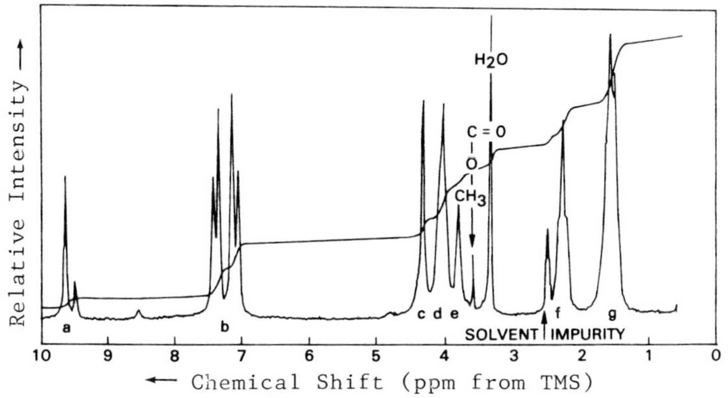

Figure 5. NMR spectrum of polymer A tensile specimen aged 30 d in methanol at 60°C.

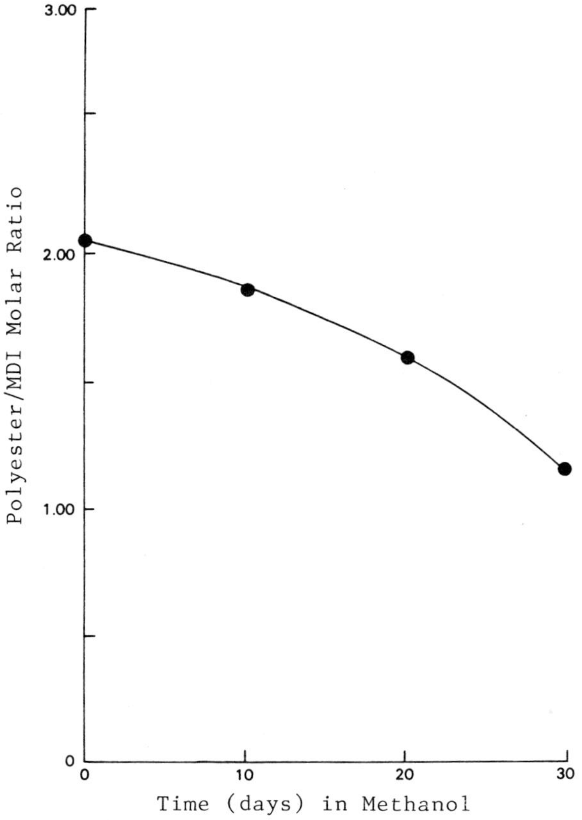

Figure 6. Relative composition of polymer A aged in methanol at 60°C.

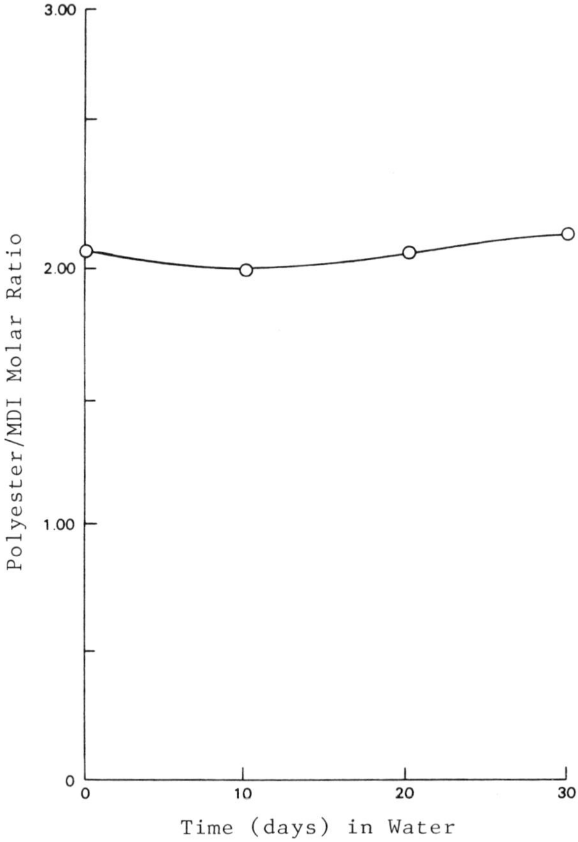

Figure 7. Relative composition of polymer A aged in water at 80°C.

deteriorated (Fig. 1). The degraded polyester is not soluble in water nor is the polyurethane highly swollen in water. Consequently, NMR techniques are not as useful for monitoring polyurethane aging in water as in methanol environments.

In general, results from NMR analysis in all media indicate that the polyester to MDI molar ratio remained roughly constant in aging media which contained water since the degraded polyester is not extracted. In addition, more change occurred in the molar ratio of polyester to MDI in polymer A than in polymer B during aging. In Table II, the amount of methyl ester (detectable by NMR analysis) is shown to increase with methanol content in isooctane solutions. This provides evidence that the extent of chain scission is proportional to methanol content.

Table II. Effect of Composition of Isooctane-Methanol Solutions on Methyl Ester Formation in Polymer B Aged 30 Days at 80°C

Methanol Content (Volume %)	Methanol Content (Mole Fraction)	Moles $-\overset{O}{\overset{\|\|}{C}}OCH_3$ / Moles Polyester
5	0.18	0.007
10	0.31	0.017
25	0.58	0.029

After the aging experiments were completed, the solvent mixtures used were evaporated to dryness. The residues were examined by proton NMR to identify the constituents lost by the tensile specimens during aging. For example, Fig. 6 indicates that polymer A loses polyester when aged in methanol. Figure 8 is the NMR spectrum of the polymer A residue. It consists almost entirely of methanol capped poly(butylene adipate). This result confirms that the decreased values of polyester to MDI molar ratios observed for the solid samples are caused by the extraction of degraded polyester from the polyurethane substrate. Observation of the methyl ester end group allows calculation of the number average molecular weight of this extracted material. For polymer A, it is ~560, whereas for polymer B aged in methanol, the residue is poly(caprolactone) with M_n~1200. The residues obtained in the isooctane-methanol (75-25) mixtures are also methanol capped polyesters.

In instances where the polyurethanes were aged in pure water, very low concentrations of the specific hydrolysis product 6-hydroxy caproic acid were observed in the residue of polymer B

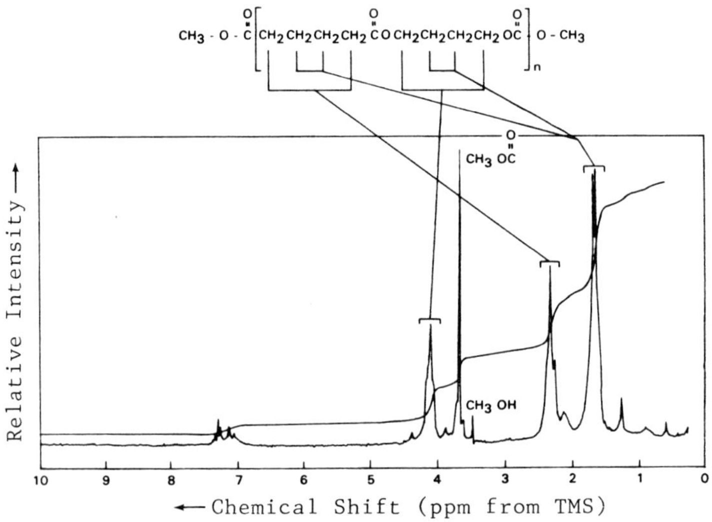

Figure 8. NMR spectrum and peak assignments of extraction residue from polymer A tensile specimen aged 30 d in methanol at 60°C ($CDCl_3$ is the solvent).

and butane diol and adipic acid were observed in the residue of polymer A. These monomeric species are all that one would expect to see dissolved in water. For the solvent mixtures of methanol-water (50-50) and isooctane-methanol-water (75-12.5-12.5), residues were rather different for polymers A and B. Both solvent systems tended to remove small amounts of polyester from polymer B, but only diols and adipic acid from polymer A.

GPC. Molecular weights were obtained by GPC. The number average molecular weights are probably slightly high because the GPC columns used did not resolve molecular weights below ~5000. A cross-plot of molecular weight, tensile strength, and aging time for polymer A in water is shown in Fig. 9. It is noted that the material can tolerate considerable molecular weight degradation before tensile properties are strongly affected. A more general plot of tensile strength with molecular weight for polymer A aged in methanol and water is shown in Fig. 10. This figure shows the expected behavior of a monotonic decrease in tensile strength with decreasing molecular weight. It also shows that molecular weight must decrease to some critical value before the precipitous drop in tensile strength begins. This plot is significant since it suggests that both water and methanol alter the molecular weight distribution in the same manner even though, as shown earlier, the degradation kinetics are vastly different (Figs. 1 and 2). In both cases, the molecular weight distribution has been altered by chain scission (in the case of water the mechanism is hydrolysis, while in the case of methanol it is transesterification). Consequently, molecular weight may be used as an independent measure of the extent of degradation.

Liquid Sorption. Liquid sorption measurements were performed gravimetrically on polymer A in methanol-water and methanol-isooctane solutions. The results are plotted in Figs. 11 and 12. In Fig. 11, a linear tie-line was drawn through the solvent uptake data points for the pure liquids. It appears that the amount of methanol (or water) absorbed by the polymer is linearly proportional to the mole fraction of methanol in the solution. However, proof of this assertion requires the independent measurement of the species dissolved in the polymer. In other words, the partitioning of the solvents within the polyurethane is not known. In contrast to the aforementioned behavior, positive deviations were observed from the linear tie-line in isooctane-methanol solutions. This indicates that an excess of one liquid, and of possibly both, was sorbed by the polymer (Fig. 12). These results also demonstrate that methanol is much more compatible with the polyurethanes than water (i.e., methanol is absorbed to a much greater degree in the polymer than water).

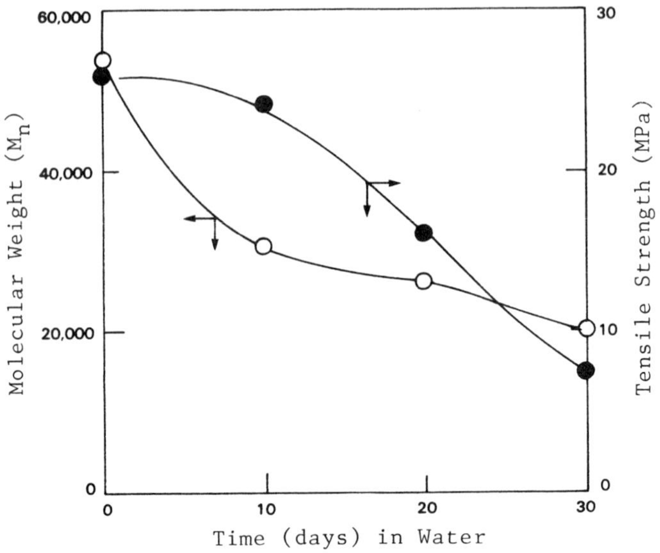

Figure 9. Effects of water on tensile strength (●) and molecular weight (○) of polymer A.

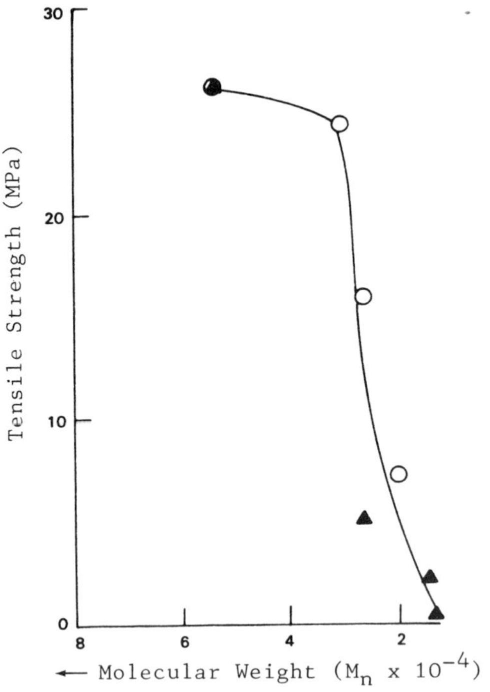

Figure 10. Dependence of tensile strength of polymer A on molecular weight. Key: ▲, aged in methanol at 60°C; and ○, aged in water at 80°C.

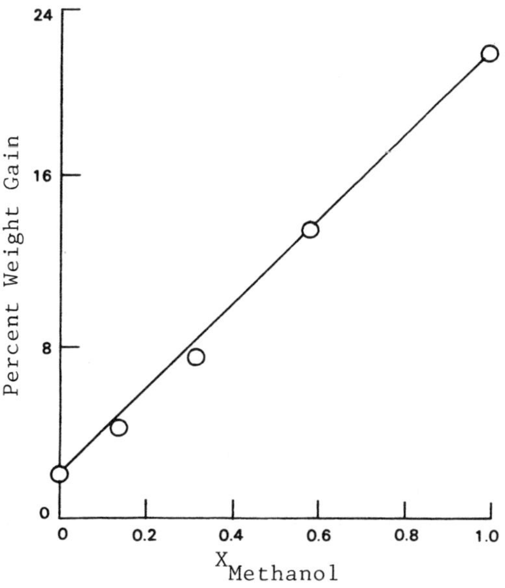

Figure 11. Liquid sorption of polymer A in methanol–water solutions at 60°C as a function of mole fraction of methanol.

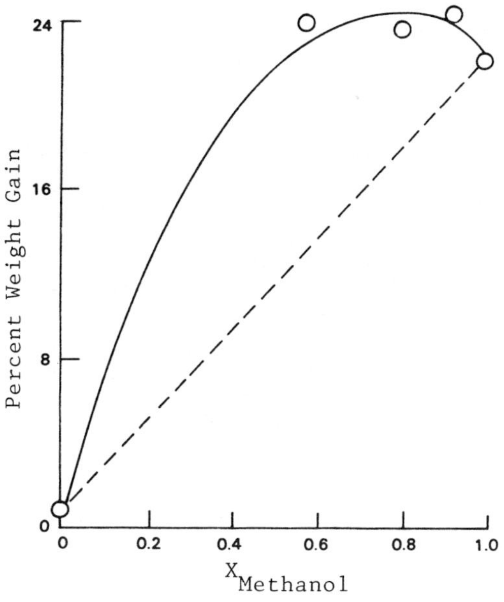

Figure 12. Liquid sorption of polymer A in methanol–isooctane solutions at 60°C as a function of mole fraction of methanol.

Discussion

The exposure of polyurethane tensile bars to various solvent mixtures, has demonstrated that both polymers A and B are very susceptible to attack by water and methanol. The loss of tensile properties occurred more rapidly in solutions containing methanol with the extent of degradation being proportional to the amount of methanol present. Although experiments have not been performed using other alcohols or glycols, available literature indicates a similar degradation would occur (3). In addition to degradation, the extensive swelling (22% weight gain) of the polyurethane by methanol is expected to adversely affect mechanical properties due to plasticization.

The NMR analysis of polymer A tensile bars aged in methanol and isooctane-methanol solutions showed a systematic decrease in the amount of polyester present with aging time. The same behavior was observed for polymer B except that the loss of polyester was less systematic. Subsequent analysis of the residue confirms that the low molecular weight polyester degradation products are dissolved into the solution in both cases.

Although the polyester/MDI molar ratio measured by NMR provides a useful quantitative measure of the extent of degradation, the results of this study indicate that molecular weight measurements are to be preferred. This is because a molecular weight decrease is a direct consequence of chain scission, whereas the loss of polyester requires both chain scission and the subsequent dissolution of polyester chain segments into the aging media. In cases where dissolution does not occur (such as was noted in pure water), the molecular weight still provides a quantitative measure of degradation. The molecular weight after aging also correlates well with physical properties which ultimately dictate the useful life of polyurethane materials in these environments.

Conclusions

1. Extensive degradation of polyester based thermoplastic polyurethane samples occurs due to hydrolysis in the presence of water and transesterification in the presence of methanol.

2. Both the hydrolysis and transesterification reactions result in a reduction in molecular weight and consequent loss of tensile properties. A good correlation is observed between molecular weight and tensile properties which permits molecular weight determinations to be used as a convenient method to monitor the extent of degradation.

Literature Cited

1. Schollenberger, C. S.; Stewart, F. D. Journal of Elastoplastics 1971, 3, 28.
2. Chokki, Y.; Nakabayashi, M.; Sumi, M. Die Makromolekular Chemie, 1972, 153, 189.
3. Penfold, J.; Kuypers, J. Proc. Urethanes and Environment Conf. H-1 (1976).

RECEIVED April 7, 1983

12

Hostile Effects of Oxygenated Solvent Interaction with Type II Rigid Poly(vinyl Chloride)

RICHARD F. MILLER

Rice University, Department of Mechanical Engineering and Materials Science, Houston, TX 77251

PAUL E. RIDER

University of Northern Iowa, Department of Chemistry, Cedar Falls, IA 50614

A hostile effect can be predicted for the interaction of various oxygenated solvents with Type II rigid poly(vinyl chloride) if the difference in the solubility parameters (δ_1-δ_2) of the solvent-polymer system is less than ± 3.2H and if the hydrogen bonding potential, HBP, of the solvent-polymer system is less than or equal to zero.

Much of our present knowledge of solutions is based on the principles developed by Hildebrand and Scott in 1924 (1). Fortunately, most organic solvents are nonpolar and therefore their intermolecular forces are weak London or dispersion forces. Hildebrand used the term "regular solutions" to describe solutions of nonelectrolytes and their nonpolar solvents. Additional theories on the solubility of polymers were developed by Flory (2) and Huggins (3). Probably the most important publications leading to the practical use of solubility theories by polymer scientists were those published by Burrell in 1955 (4) and 1966 (5). Modifications in the Hildebrand solubility parameter concept for regular solutions to account for larger intermolecular forces were made by Liebermann (6), Crowley (7), Hansen and Beerbower (8) and Nelson et al. (9).

To evaluate the potential hostile effects of oxygenated solvents on Type II rigid poly(vinyl chloride), one must consider that PVC has very limited solubility. The most effective solvents are those which appear to be capable of some form of interaction with the polymer. Small (10) suggested that PVC is a weak proton donor and effective solvents are proton acceptors. Thus it was proposed that the polymer would be soluble in tetrahydrofuran; ketones, e.g. cyclohexanone, methyl isobutyl ketone and nitro compounds such as nitrobenzene. Hansen (11) evaluated the solvency effects of twenty-six solvents on poly(vinyl chloride), a summary of his data with respect to oxygenated solvents is presented in part in Table I.

0097-6156/83/0229-0193$06.00/0
© 1983 American Chemical Society

Table I. Solubility Data for Poly(vinyl chloride) with Various Oxygenated Solvents (11-13, 21)

Solvent	δ	b	C	HBP	PVC Solubility
Methanol	14.5	1.80	1.68	0.35	I
Ethanol	10.0	1.70	1.74	0.32	I
1-Propanol	10.5	1.60	1.80	0.28	I
1-Butanol	13.6	1.60	1.83	0.29	I
1-Pentanol	11.6	1.60	1.81	0.28	I
Cyclohexanol	10.6	1.60	1.86	0.30	I
Acetone	9.9	0.50	1.74	-0.45	M
Methyl Ethyl Ketone	9.3	0.34	1.70	-0.52	M
Methyl Isobutyl Ketone	8.58	0.31	1.65	-0.49	M
Cyclohexanone	9.9	0.43	1.80	-0.54	S
Isophorone	9.9	1.05	1.85	-0.11	M
Ethyl Acetate	9.0	0.75	1.60	-0.22	I
Propyl Acetate	8.8	(0.80)	1.60	(-0.20)	M
n-Butyl Acetate	8.65	0.85	1.60	-0.18	M
Isoamyl Acetate	7.8	0.89	1.47	-0.01	M
Diethyl Ether	7.4	1.20	1.96	0.00	I
Dioxane	10.0	1.20	1.85	0.00	M
Tetrahydrofuran	9.5	0.73	1.70	-0.28	S
Methyl-t-Butyl Ether	7.07	0.965	1.96	-0.20	M
Dimethylformamide	12.1	1.11	2.10	-0.09	M
Dimethyl Sulfoxide	12.0	0.77	2.32	-1.07	M

S-completely soluble; M-some visible interaction; I-totally insoluble

( )-values given are estimated averages

Since Hansen's work was completed, Rider (12, 13) has developed both a two-parameter and a four-parameter model in order to estimate the effects of the hydrogen bonding potential, HBP, according to the equation shown:

$$HBP = (b_1 - b_2)(C_1 - C_2)$$

wherein, the donor is assigned a parameter, b, that reflects its donating tendency and the acceptor is assigned a parameter, C, that reflects it accepting tendency. The b parameter has units of Kcal/mole and the C parameter is dimensionless.

For the vinyl chloride resin, the monomeric unit is ethyl chloride, therefore, the value for b is estimated to be approximately 1.2 Kcal/mole (12) while the value for C of 1.10 was taken from the observed frequency shift of methanol bonded chloroform as measured by Crowley et al. (7) and from the b value and slope for the linear enthalpy - frequency shift relationship reported by Rider (12, 13).

Therefore, according to the Rider data (12, 13) if the HBP is positive, this represents unfavorable hydrogen bonding and suggests that solutions would not be formed. A negative HBP value would reflect favorable hydrogen bonding, thereby suggesting solution formation. In practice, Seymour (14) states that as a first approximation, and in the absence of strong interactions such as hydrogen bonding, solubility or compatability on the molecular level can be expected if $\delta_1 - \delta_2$ is less than ± 1.8 H. The question now arises whether hostile effects include the action wherein the physical and/or chemical integrity of a polymer is affected without the resultant formation of a solution. In this particular study, the formation of "regular solutions" will not be considered. Instead the possible hostile effects of oxygenated solvents will be discussed wherein the solvents investigated were expected to exhibit no or as yet unreported interactions with poly(vinyl chloride).

Semi-rigid or Type II rigid poly(vinyl chloride) specimens described by Seymour (15) were initially investigated to determine if such polymers could be used in service where the PVC could come into contact with either t-butyl alcohol (TBA) or methyl t-butyl ether (MTBE). This work now has been expanded to include a variety of oxygenated solvents (Table I) including those studied by Hansen (11).

EXPERIMENTAL

To evaluate the hostile effects of various oxygenated solvents on PVC, static immersion tests were conducted wherein test specimens having an approximate surface area of 19.4 cm^2 and an edge thickness of approximately 0.381 cm were exposed to from 20 to 200 ml of solvent. The tests were run at 25°C in glass test chambers having glass racks on which the coupons were suspended. Exposure times ranged from 6 to 72 hours, the coupons were then removed, dried and reweighed in the initial studies involving t-butyl alcohol and methyl t-butyl ether (Table II).

Table II. Weight Changes of PVC Specimens Exposed to t-Butyl Alcohol and Methyl t-Butyl Ether

Solvent	Time (hrs.)	wt. change (mg)	% wt. change
Blank	6	0.9	0.08
	72	1.4	0.12
t-Butyl Alcohol	6	- 0.5	-0.05
	72	- 0.7	-0.06
Methyl t-Butyl	6	- 9.5	-0.81
Ether	72	-19.0	-1.62

In the various other solvents employed the specimens were visually inspected and submitted for physical testing in an effort to discern changes in tensile or pulling strength according to a modification of the procedure reported by Schnell (16). These results are shown in Table III.

Table III. Tensile and Ultimate Strengths of PVC Specimens Exposed to a Hostile Enviroment

Solvent	Ultimate Strength (kg)	Tensile Strength 10^6 kg/m^2
Blank	55.4	4.29
t-Butyl Alcohol	56.3/56.75	4.43
Methyl t-Butyl Ether	43.1/41.31	3.45/3.23
2-Propanol	55.8	4.37
Propyl Acetate	43.58	2.97
Acetone	25.4	1.47
Dioxane	44.49	2.57
Dimethylformamide	disintegrated	
Methyl Ethyl Ketone	disintegrated	

Attenuated total reflectance infrared spectroscopy was employed to determine the possible chemical modification of the PVC specimen exposed to t-butyl alcohol and methyl t-butyl ether. Infrared spectroscopy has been used to study solvent absorption (17), oxidation (18) and other degradation reactions of polymers (19). In the studies of the hostile effects of methyl t-butyl ether and acetone, the solvent was concentrated and examined using conventional infrared techniques.

Finally, the PVC specimen exposed to methyl t-butyl ether (MTBE) was examined by scanning electron microscopy in an effort to evaluate changes in the surface characteristics of the polymer.

Results and Discussion

When reviewing the spectrum obtained on the PVC test specimen (Figure 1), the outstanding absorptions attributable to the addition of Cl appear at 8μ (1250 cm^{-1}), CH_2 deformation, and at 15μ (665 cm^{-1}), carbon-halogen stretch. These absorptions gradually weaken as the chain length increase, however, when Cl is not terminal, the stretch absorptions, which appear practically without exception at 13.8 μ and 15.5 μ (725 and 645 cm^{-1}) for chlorine, shift beyond 16 μ (625 cm^{-1}). It is difficult to assign specific characteristic absorptions in this particular spectrum when several halogen atoms are introduced into any one molecule because of the broad range in which the C-Cl stretch bands appear. The I.R. spectrum is further complicated by the presence of plasticizers and fillers.

When considering the spectrum obtained after the PVC specimen was exposed to MTBE (Figure 2), a rather significant absorption appears at 610 cm^{-1} which is not characteristic of a C-Cl stretch band. It appears that this rather strong absorption results from the presence of TiO_2 on the surface of the test specimen. Initially it was believed that the MTBE was able to dissolve the PVC uniformly thereby leaving a coating of TiO on the surface of the specimen, however, the reaction of a polymer with a corrosive is dependent upon diffusion of the attacking reactant. If the polymer is at a temperature above its T_g there is a constant segmental motion that favors diffusion (20). To better understand the possible cause of the apparent hostile effect, the MTBE solution was concentrated under vacuum and analyzed. Infrared analysis of the concentrated MTBE solution revealed the presence of dioctyl phthalate, a plasticizer, and a lesser amount of a carboxylate. Therefore, it is reasonable to assume that the MTBE is compatible with the PVC and thus able to diffuse into and out of the polymer carrying with it the plasticizer, dioctyl phthalate, and TiO_2. However, because TiO_2 is insoluble in the MTBE it was deposited on the specimen's surface while the various organics were solubilized.

Assuming MTBE was able to diffuse into and out of the PVC, carrying with it the various organic additives and depositing TiO on the surface of the specimen, then one would expect to see some surface adulteration when compared to a specimen which was not exposed to such a hostile environment.

As expected, a rather dramatic difference was seen when comparing the surface of the specimen exposed to MTBE to one which had not (Figure 3,4). A significant "smoothing" of the surface of the specimen exposed to MTBE tends to support the

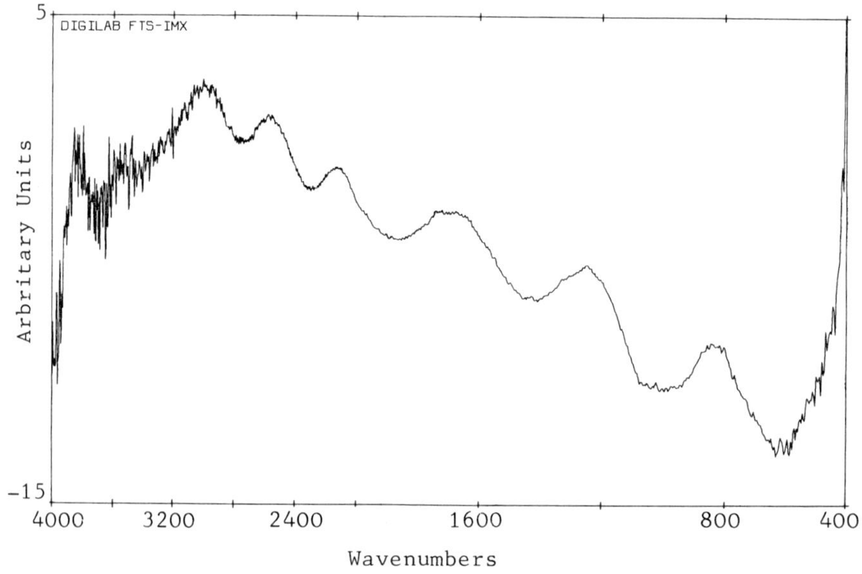

Figure 1. IR spectrum of PVC test specimen.

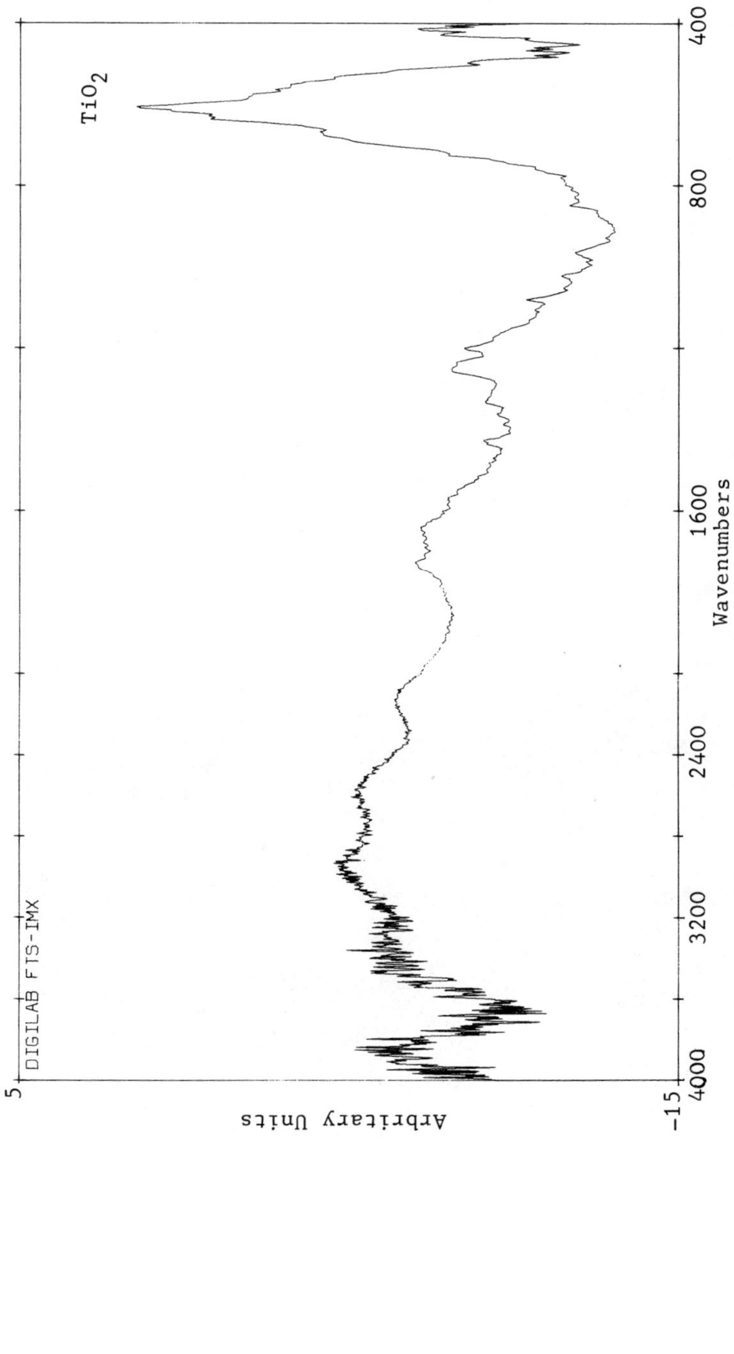

Figure 2. IR spectrum of PVC exposed to MTBE.

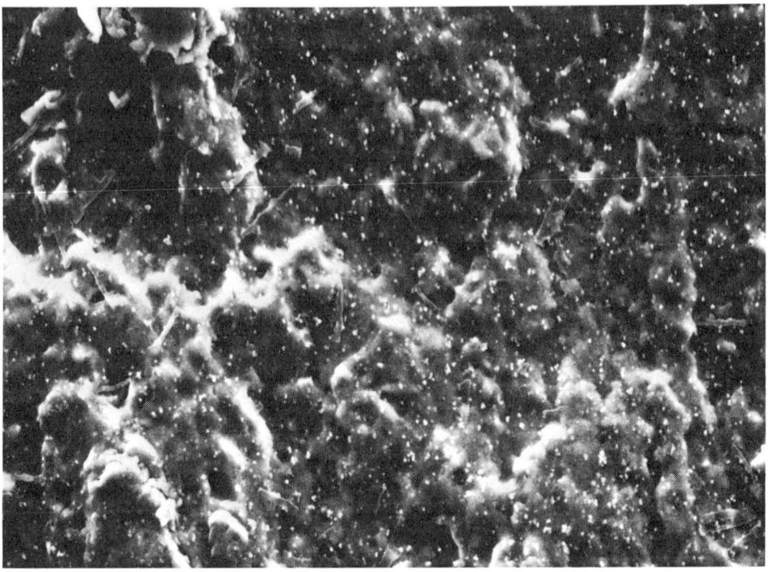

Figure 3. Scanning electron micrograph of PVC test specimen (×2000).

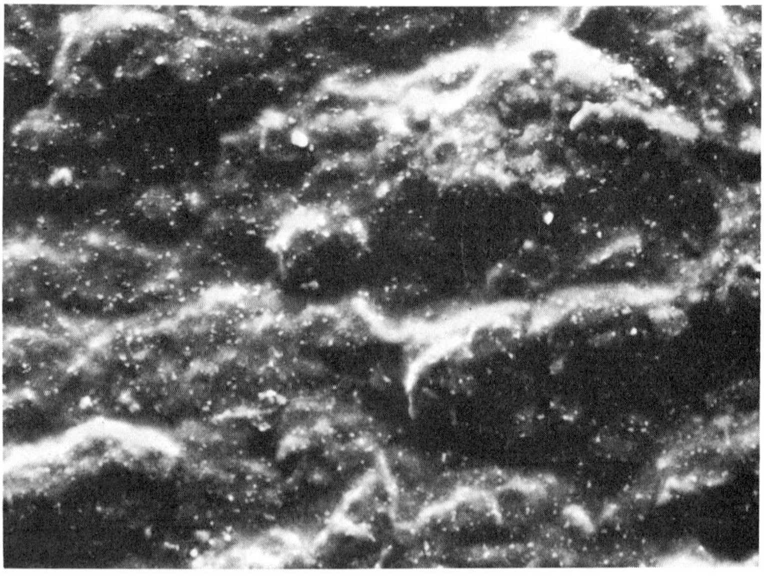

Figure 4. Scanning electron micrograph of PVC exposed to MTBE (×2000).

hypothesis that the MTBE aided in the deposition of a rather uniform layer of TiO_2 on the surface of the specimen as a result of diffusion.

The acetone employed in another static immersion test was concentrated as in the case of the MTBE and analyzed. The infrared spectrum of the acetone concentrate indicated the presence of a carboxylic acid and possibly a short chain alcohol species. Comparison of the infrared spectrum obtained to those of known polymer binding agents indicated that the carboxylic acid is similar to those in the Shell Versatic acid series, closely resembling Versatic 911 (22).

Subsequent NMR analysis of the acetone concentrate showed the presence of a multi-hydroxyl containing compound such as pentaerythritol and at least two types of ether moeities, O-methyl and $-OCR_3$. Although the structure of the acetone soluble organics has not been completely elucidated, it is interesting to note the differences between the MTBE solubles and the acetone solubles, both of which exhibit hostile effects toward PVC. It appears that MTBE, $\delta = 7.07$, extracts less polar substituents whereas acetone, $\delta = 9.9$, extracts the more polar binding agents. The absence of an aromatic absorption in both the infrared and the NMR spectrum of the acetone soluble fraction tends to support this conclusion.

The data presented in Table III show the effect of hostile solvents on the tensile strength of PVC. Although numerous other oxygenated solvents were employed a substantial number of the test specimens essentially disintegrated while attempting to remove them from the test chamber. However, with the data presented, one can easily see that a hostile effect adversely affects both the chemical and physical integrity of the polymer.

The definition of a hostile effect, proposed earlier, as an adverse interaction between polymer and solvent wherein the chemical and/or physical integrity of the polymer is affected without the resultant formation of a "regular solution" has been clearly shown. The need now arises for an improved method by which hostile effects can be predicted. Therefore, if one employs the generalized rule cited by Seymour (14) that compatability or solubility on the molecular level can be expected if $\delta_1 - \delta_2$ is less than $\pm$ 1.8H and that the swelling of polymers occurs when $\delta_1 - \delta_2$ is equal to $\pm$ 3.2H (20), then one would expect some interaction between PVC and all of the solvents listed on Table I, with the possible exception of methanol and 1-butanol.

If the data presented by Rider (12, 13) is considered then one would expect a solvent/polymer interaction with all of the previously mentioned solvents with the exception of the alcohols and possibly dioxane. However, if one considers the generalized rule of Seymour (14, 20) and the hydrogen bonding potential data presented by Rider (12, 13) a suitable method by which a hostile effect can be predicted is achieved. Simply stated, a hostile effect can be predicted if:

(1) the (δ_1 - δ_2) of the solvent-polymer system is less than ± 3.2H

and

(2) the hydrogen bonding potential of the solvent-polymer system is less than or equal to zero

Application of this theory to the systems investigated is consistent with the data obtained and the results are shown in Table IV.

Table IV. Hostile Effect Data for Poly(vinyl chloride)

Solvent	δ_1-δ_2	HBP	Hostile
Methanol	5.35	0.35	no
Ethanol	0.85	0.32	no
1-Propanol	1.35	0.28	no
t-Butanol	3.25	(0.25)	no
Acetone			
Methyl Ethyl Ketone	0.15	-0.52	yes
Methyl Isobutyl Ketone	-0.57	-0.49	yes
Cyclohexanone	0.75	-0.54	Soluble
Ethyl Acetate	-0.15	-0.22	yes
n-Propyl Acetate	-0.35	(-0.20)	yes
n-Butyl Acetate	-0.50	-0.18	yes
Dioxane	0.85	0.00	yes
Tetrahydrofuran	0.35	-0.28	Soluble
Methyl-t-Butyl Ether	-2.08	(-0.20)	yes
Dimethylformamide	2.95	-0.09	yes
Dimethyl Sulfoxide	2.85	-1.07	yes

( ) indicates values that are estimated based on available data

Literature Cited

1. Hildebrand, J.H. and Scott, R.L., "The Solubility of Nonelectrolytes", 3rd ed., Reinhold Corporation, New York, 1950.
2. Flory, P.J., J. Chem. Phys., 1941, 9, 660.
3. Huggins, M.L., J. Chem. Phys., 1941, 9, 440.
4. Burrell, H., Official Digest, Fed. Soc. Paint Technol., 1955, 27, 726.
5. Burrell, H., Official Digest, Fed. Soc. Paint Technol., 1966, 38, 34.
6. Liebermann, E.P., Official Digest, Fed. Soc. Paint Technol., 1962, 34, 444.
7. Crowley, J.D., Teague, G.S. and Lowe, J.W., J. Paint Technol., 1966, 38, 269.
8. Hansen, C.M. and Beerbower, A., "Encyclopedia of Chemical Technology", John Wiley and Sons, New York, 2nd ed., 1971, p. 889.
9. Nelson, R.C., Hemwall, R.M. and Edwards, G.D., J. Paint Technol., 1970, 42, 636.
10. Small, P.A., J. Appl. Chem., 1953, 3, 71.
11. Hansen, C.M., J. Paint Technol., 1967, 39, 104.
12. Rider, P.E., J. Appl. Polymer Sci., 1980, 25, 2975.
13. Rider, P.E., "Hydrogen Bonding Potential, HBP: A Model for Predicting Resin Solubilities in Organic Solvents", submitted for publication, 1982.
14. Seymour, R.B. and Carraher, C.E., "Polymer Chemistry", Marcel Dekker, Inc., New York, 1981, pp. 61-68.
15. Seymour, R.B., "Modern Plastics Technology", Reston Publishing Company, Inc., Reston, Va., 1975, pp. 192-193.
16. Schnell, H., Inst. Eng. Chem., 1959, 51, 157.
17. Foster, R.S. and Vokening, V.B., Corrosion, 1959, 15(10), 85.
18. Seymour, R.B., Tsang, H.S., and Warren, D., Poly. Eng. Sci., 1967, 7, 55.
19. Jellinck, H.H.G., Flausmann, G.F., and Kryman, F.J., J. Appl. Polymer Sci., 1969, 13, 107.
20. Seymour, R.B., Plastics vs. Corrosives, John Wiley and Sons, New York, 1982, p. 46.
21. Seymour, R.B., "CRC Handbook of Chemistry and Physics", CRC Press, Inc., Cleveland, 61st ed., 1980, pp. 687-690.
22. Afremov, L.C., Isakson, K.E., Netzel, D.A., Tessari, D.J. and Vandeberg, J.T., eds., "Infrared Spectroscopy: Its Use in the Coatings Industry", Federation of Societies for Paint Techology, Philadelphia, 1969, pp. 137-155.

RECEIVED February 22, 1983

13

Chemical and Solvent Resistance of Selected Styrene–Maleic Anhydride Copolymers

RICHARD F. MILLER

Rice University, Department of Mechanical Engineering and Materials Science, Houston, TX 77251

The chemical and solvent resistance of selected styrene-maleic anhydride copolymers was investigated and their response compared to oriented polystyrene. It was found that the maleic anhydride moeity appears to increase the reactivity of SMA copolymers toward oxidizing acids while glacial acetic and hydrochloric acids hydrolyze esters of SMA copolymers. The solvent resistant properties of SMA copolymers compare favorably to those of polystyrene.

In 1930, Wagner-Jauregg described alternating copolymers of styrene and maleic anhydride (1). The tendency for regular alternation instead of a random sequence of styrene and maleic anhydride may be explained by the low reactivity ratio of maleic anhydride (r_1) which is approximately 0 and the higher reactivity ratio of styrene (r_2) which is approximately 0.0095 at 25°C (2). Thus, a value of 0 for the product r_1r_2 is considered as a criterion for alternating copolymers. However, both block copolymers (3-6) and random copolymers of this system have also been described (7-12).

The formation of block copolymers can result from the addition of excess styrene monomer to SMA macroradicals (13). Maleic anhydride has also been reported to homopolymerize when initiated by gamma-radiation of free radical initiators. The highest conversions were obtained employing acetic anhydride as the solvent, in a ratio of solvent to monomer of 75:25 (14,15).

Alternating copolymerization is dependent upon the presence of a charge transfer complex which serves as the polymerizable species (16,17). Since the stability of the chain transfer complex decreases as the temperature is increased, the tendency for alteration decreases at elevated temperatures. Thus, random copolymers are produced when the polymerization occurs at temperatures above the threshold limit for the existence of the charge transfer complex (18).

0097–6156/83/0229–0205$06.00/0
© 1983 American Chemical Society

SMA copolymers have been investigated in a number of applications. A 2:1 molar styrene-maleic anhydride copolymer can be added to commercial injection moldable polystyrene to give permanent electrostatic protection (19). Crosslinking of a half acid/ester derivative of a styrene-maleic anhydride copolymer with epoxy resins, catalyzed by zinc octanoate, yields films of high gloss and excellent chemical resistance (20). The addition of a 1:1 molar styrene-maleic anhydride copolymer to a xylene-formaldehye resin when reacted with epoxies gives superior casting resins. The reaction product of a styrene-maleic anhydride copolymer and N,N-dimethylpropyldiamine which was quaternized with methyl iodide, was found to be an excellent, long lasting oil slick dispersant (22).

The behavior and durability of synthetic resins in hostile enviroments has become an important consideration. Therefore, this investigation will focus on the chemical and solvent resistant properties of selected styrene-maleic anhydride copolymers.

Experimental

Commerically available samples of biaxially oriented polystyrene and SMA copolymer sheet material, having a thickness of 0.0381cm, were used in this investigation. It is generally recognized that crystallization under stress can enhance the tensile properties of a semi-crystalline polymer through a special arrangement of the crystalline portion (23). Therefore, the physical properties of the styrene-maleic anhydride copolymers chosen were compared to those of polystyrene produced in the same manner and are shown in Table I (24).

Table I. Physical Properties (24)

Property	OPS	SMA	SMA (ester)
Tensile Strength, 10 kg/m	5.6	3.37	3.59
Tensile Elongation, %	3-40%	20%	10%
Tensile Modulus, 10 kg/m	3.17	2.26	2.32
Rockwell Hardness	85 M	80 L	77 L
Specific Gravity,	1.05	1.06	1.08

Attenuated total reflectance infrared spectra were obtained on the oriented polystyrene sample (Figure 1) and on the two SMA copolymers selected for investigation (Figure 2,3).

Carbon, hydrogen and oxygen analyses were obtained on the SMA copolymers. Values obtained were C 87.79%, H 7.05% and O 4.81% and C 89.14%, H 7.68% and O 2.81%.

To determine the effects of exposure to hostile enviroments, such as strong oxidizing acids, coupons of the polymers were immersed for seven days at ambient temperatures in concentrated nitric acid and 10% nitric acid. These coupons were also exposed to glacial acetic acid and hydrochloric acid according to ASTM procedure D-543-67. Visual observations were made upon removing the coupons from the acid enviroment (Table II) and the coupons were examined by means of attenuated total reflectance infrared spectroscopy which has been used to study solvent absorption (25), oxidation (26) and other degradation reactions of polymers (27).

Table II. Chemical Resistance of OPS and Styrene-maleic Anhydride Copolymers to Strong Oxidizing Acids (Visual Appraisal after 7 days immersion at ambient temperature) ASTM-D-543-67

Exposure Media	OPS	SMA	SMA (ester)
Nitric (Conc.)	yellowed, flexible	yellow, brittle	yellow, brittle
Nitric 10%	no apparent	no apparent	no apparent
Hydrochloric (Conc.)	slightly whitened	whitened, crazed	whitened, crazed
Acetic (Glacial)	no apparent interaction	whitened, crazed	whitened, crazed

The effects of various oxygenated solvents were determined by employing a static immersion test wherein a test specimen having the approximate surface area of 6.45 cm^2 and an edge thickness of 0.0381 cm was exposed to from 20 to 200 ml of solvent. The results of exposure to the various oxygenated solvents are reported in Table III along with the solubility parameter of each solvent (28), the hydrogen bonding donor parameter, b, the hydrogen bonding acceptor parameter, C and the hydrogen bonding potential, HBP (29,30).

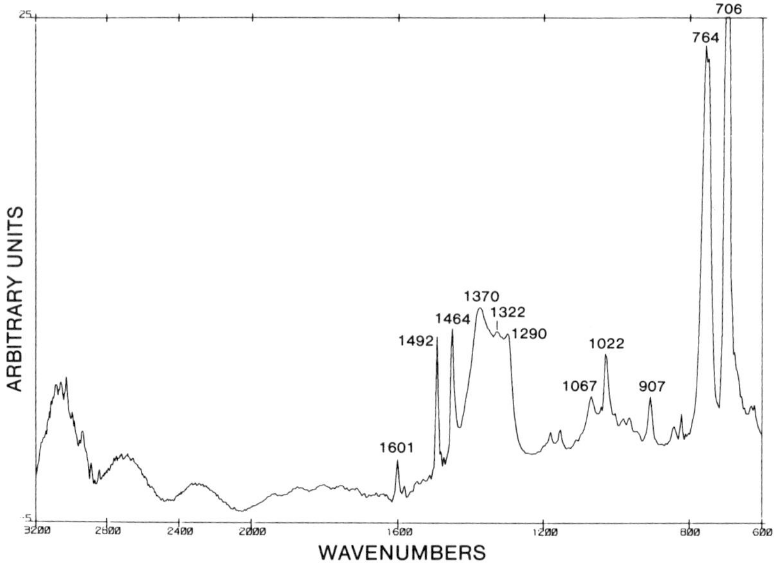

Figure 1. IR spectrum of the polystyrene test specimen.

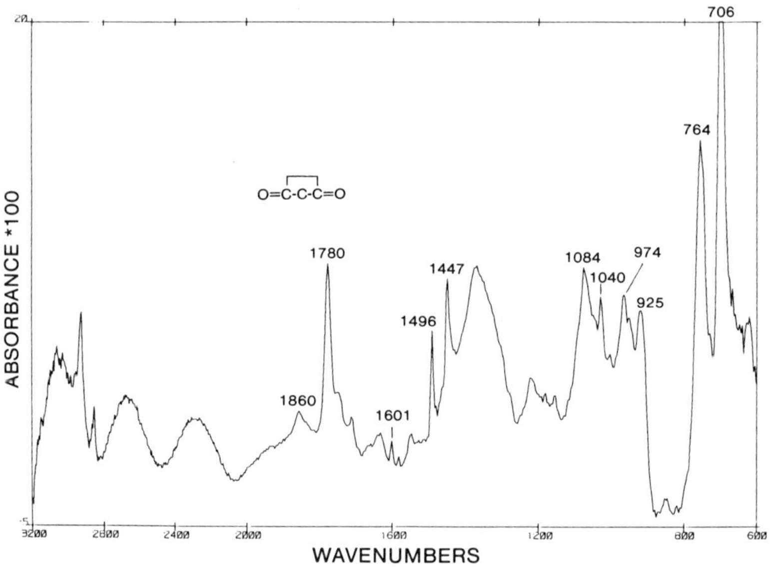

Figure 2. IR spectrum of the styrene–maleic anhydride test specimen.

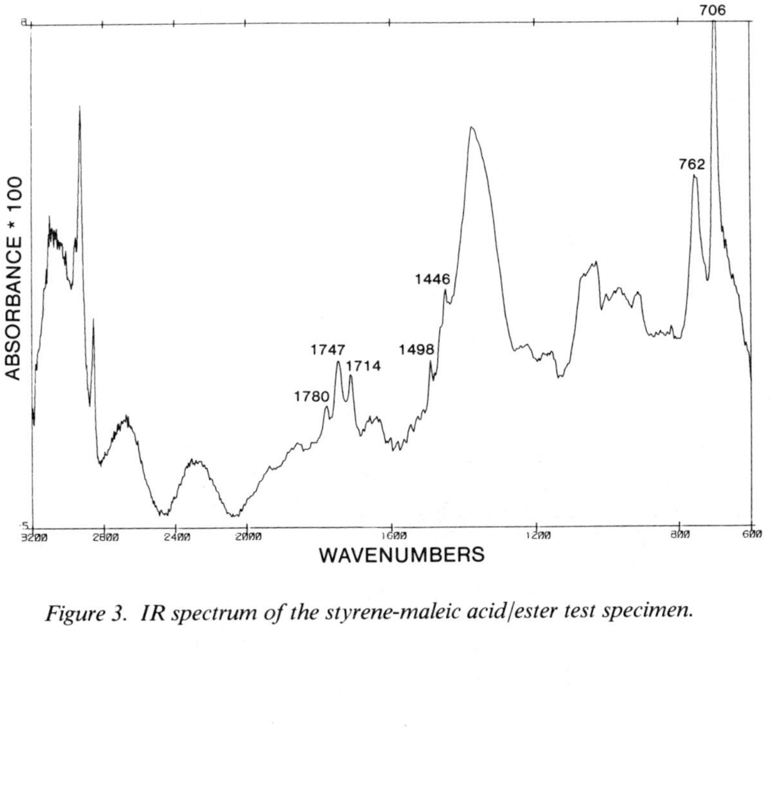

Figure 3. IR spectrum of the styrene-maleic acid/ester test specimen.

Table III. The Solvent Resistance of Selected Styrene-Maleic Anhydride Copolymers To Various Oxygenated Solvents as Compared to Polystyrene (36)

Corrosive	δ	b	C	HBP	SMA	SMA	SMA_2
Methanol	14.5	1.80	1.68	0.35	S	Q	S
Ethanol	10.0	1.70	1.74	0.32	S	Q	S
1-Propanol	10.5	1.60	1.80	0.28	-	S	S
1-Butanol	13.6	1.60	1.83	0.29	Q	S	S
1-Pentanol	11.6	1.60	1.81	0.28	-	S	S
Benzyl Alcohol	11.4				U	U	U
Acetone	9.9	0.50	1.74	-0.45	Q	U	U
Methyl Ethyl Ketone	9.3	0.34	1.70	-0.52	U	U	U
Methyl Isobutyl Ketone	8.53	0.31	1.70	-0.49	U	U	U
3-Methyl-2-butanone	8.85	(0.35)	(1.64)	(-0.49)	-	U	U
Methyl Formate	10.2				U	U	U
Ethyl Acetate	9.0	0.75	1.60	-0.22	U	U	U
Propyl Acetate	8.8	(0.80)	1.60	(-0.20)	U	U	U
Methyl-t-Butyl Ether	7.07	(0.97)	(1.96)	(-0.20)	-	U	U
Dioxane	10.0	1.20	1.85	0.00	U	U	U
Tetrahydrofuran	9.5	0.73	1.70	-0.28	-	U	U
Dimethylform amide	12.1	1.11	2.10	-0.09	U	U	U
Dimethyl Sulfoxide	12.0	0.77	2.32	-1.07	U	U	U

S - Satisfactory: Q - Questionable U - Unsatisfactory:

The solubililty parameters of the SMA copolymers selected for this study were calculated using the Small equation (31) or by titrating with more polar and less polar solvents. In this method, 1.0 ± 0.02 g of polymer was dissolved in a known volume of toluene and a measured volume of methanol was added until a stable turbid system was produced. The procedure was repeated with n-heptane. The solubility parameter of the copolymer was then determined from the summation of the product of the volume of toluene and its solubility parameter and that of non-solvent and its solubility parameter. The average value of these titrations was used to obtain the upper and lower turbidity limits and the mean of these two values was used as the solubility parameter (32).

Results and Discussion

The infrared spectrum of the polystyrene sample investigated (Figure 1) was equivalent to that of published standards (33).

The spectrum of one of the SMA copolymers investigated (Figure 2) is also consistent with that of published spectra (33). The strong absorption band at approximately 706 - 764 cm^{-1} are indicative of aromatic rings containing five adjacent hydrogen atoms, while the absorption bands at 1780 and 1860 cm^{-1} are consistent with those of a saturated five membered cyclic anhydride. The two shoulders on the 1780 cm^{-1} peak possibly indicate some hydrolysis as the result of processing.

The carbon, hydrogen, oxygen analyses obtained indicate that the polymer is approximately a 90% styrene/10% maleic anhydride copolymer wherein theoretical values for C, H and O are 87.98%, 6.92% and 4.89% versus values of 87.79%, 7.05%, and 4.81% found for C, H and O, respectively. This conclusion is further supported by the solubility parameter of the copolymer which was found to be equal to 9.65 H.

The infrared spectrum of the SMA copolymer labeled SMA-2, is consistent with other SMA copolymer spectra published (33). However, the absorption peaks in the range of 1700 - 1820 cm indicate that this copolymer has been partially esterified to yield a half/acid ester of maleic anhydride which should exhibit peaks in the range of 1700 - 1725 cm^{-1} and 1735 -1750 cm (Figure 3). The small absorption bands at 1780 and 1820 cm indicate the presence of a small amount of unreacted maleic anhydride. These data appear to be consistent for those of a styrene-maleic anhydride copolymer reported by Muskat (34). The carbon, hydrogen, oxygen analyses indicate that this polymer is a half/acid ester of a 90% styrene/10% maleic anhydride copolymer wherein theoretical values of C, H and O are 88.89%, 7.01% and 2.58% versus values of 89.14%, 7.68% and 2.81% found for C, H and O, respectively. The solubility parameter was found to be equal to 9.47 H.

The infrared spectrum of the polystyrene sample exposed to concentrated nitric acid (Figure 4) showed a decrease in absorption in the range of 1250 - 1400 cm^{-1}, a slight shift in the previously observed peaks in the range of approximately 820 and 860 cm^{-1}, and a slightly enhanced absorption at 1540 cm^{-1}. These changes are usually associated with the nitration of an aromatic ring. The aromatic nitro group usually inhibits two strong bands near 6.5 μ(1540 cm^{-1}), asymmetric stretch, and 7.4μ (1350 cm^{-1}), symmetric stretch. The 7.4 μ(1350 cm^{-1}), band is generally the weaker, however, this is not clearly evident because of the significant absorption in this range attributed to the methylene C-H bend.

The appearance of the 1540 cm^{-1} band and the shift of the absorptions of the range of 820-860 cm^{-1}, suggests that the nitric acid attacks the aromatic ring in the polystyrene chain, however, this attack appears to be very slight which is supported by the observed appearance and retained flexibility of the sample.

Upon exposure to concentrated nitric acid, the SMA copolymer (Figure 5) exhibits dramatic changes in the infrared absorption spectrum. The intensities of the two absorption bands at 706 and 764 cm^{-1} are greatly reduced while the peak observed at 764 cm^{-1} is shifted to about 750 cm^{-1}. Two strong absorption bands observed at approximately 822 and 856 cm^{-1}, indicate the presence of aromatic rings having only two adjacent hydrogen atoms, thereby suggesting substitution in the para-position. The presence of the peak a 1516 cm^{-1} is further evidence of ring substitution with its shift being explained because of conjugation. The strong absorption bands at approximately 1300, 1354 and 1380 cm^{-1} suggest that the polymer backbone is essentially unaffected, the band absorptions being enhanced by the presence of the additional electron withdrawing nitro group.

The SMA half/acid ester, when exposed to concentrated nitric acid (Figure 6), again exhibits dramatic changes in the infrared spectrum. Based on the previous discussion, it is again suggested by the data that the nitric acid attacks the para-position of the aromatic ring but to a much lesser degree than seen in the previous SMA spectrum (Figure 5). Relative reactivities toward concentrated nitric acid are obtained by ratioing the bands located in the range of 1520 - 1549 cm^{-1}.

reactivity:	SMA:	SMA-partially esterified:	PS
relative ratios:	4.39:	1.73:	1

It is therefore apparent that the SMA copolymer exhibits a higher degree of reactivity toward oxidizing acids. Partially esterifying the maleic anhydride significantly reduces this ractivity as a result of decreasing the electron withdrawing effects of the maleic anhydride on adjacent aromatic rings.

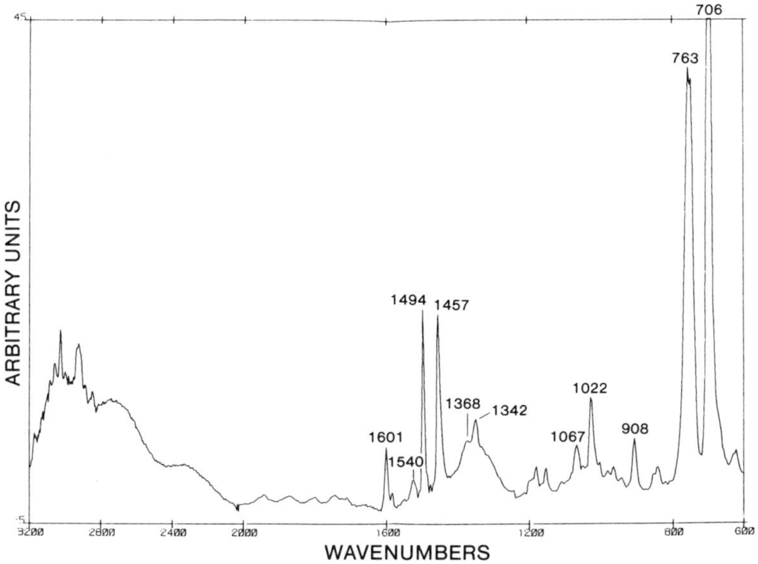

Figure 4. IR spectrum of polystyrene exposed to nitric acid.

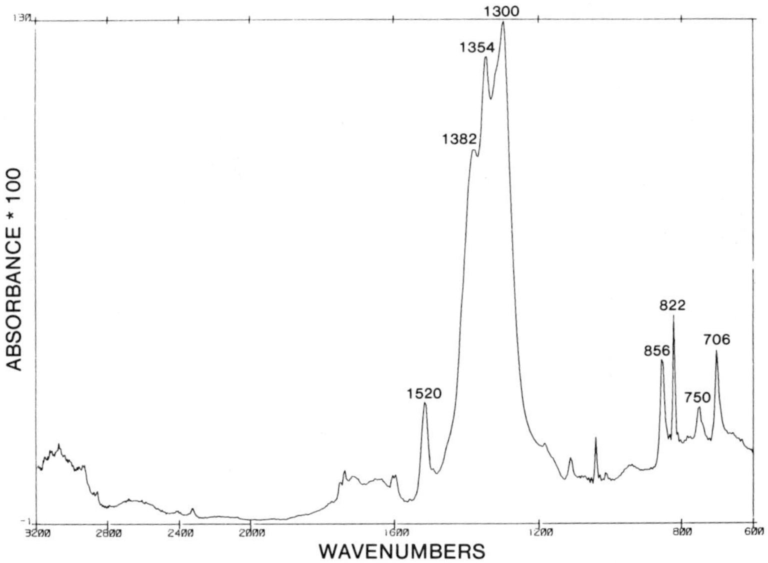

Figure 5. IR spectrum of the styrene–maleic anhydride copolymer exposed to nitric acid.

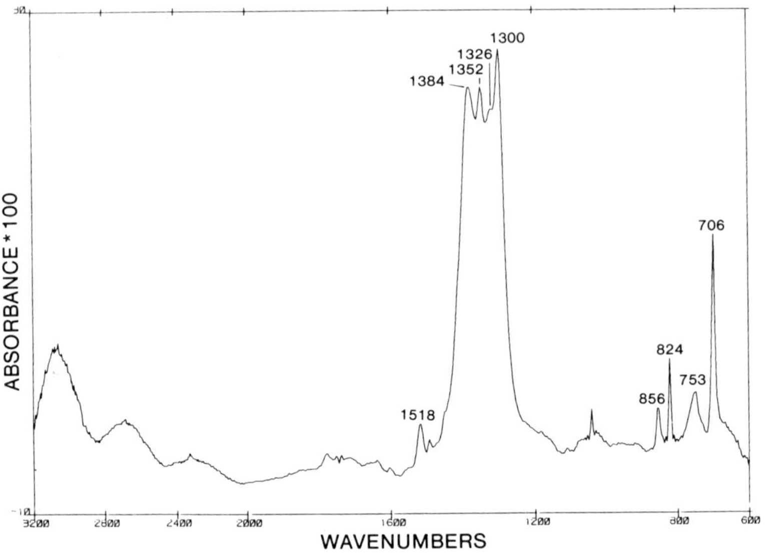

Figure 6. IR spectrum of the styrene–maleic acid/ester exposed to nitric acid.

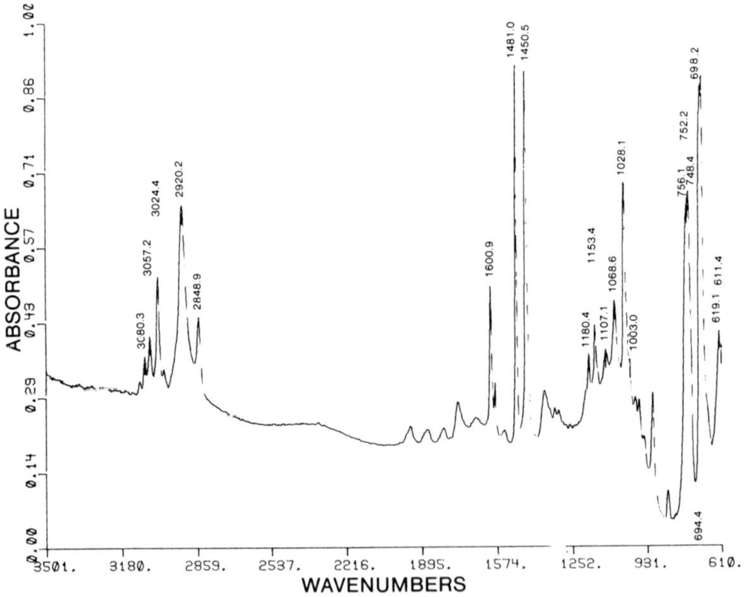

Figure 7. IR spectrum of polystyrene exposed to hydrochloric acid.

However, the position of substitution cannot be assured without additional spectral data because the substitutional bands between 11 and 15 μ(910 - 665 cm^{-1}) are strongly altered by the presence of the nitro group on the ring.

The infrared spectrum of the polystyrene exposed to concentrated hydrochloric acid (Figure 7) exhibits six peaks in the region of 690-756 cm^{-1}. The substitution patterns between 11 and 15μ (910-665 cm^{-1}) do, for the most part, follow those observed for aromatic hydrocarbons except that the band around 14μ (715 cm^{-1}) caused by dissimilar groups opposite each other on the benzene ring does not always follow the rules, therefore, a definite assignment of peaks in the range is not possible. The othe aromatic halogen bands between 8.5 and 10 μ (1175 -1000 cm^{-1}) are also difficult to assign due to other groups absorbing in this range. As a result, a comparison was made betwen the acquired spectrum and those of polystyrene (Figure 1) and a standard spectrum of poly (4-chlorostyrene). A peak table is shown in Table IV.

As seen in the peak table presented, the shifts in the aromatic absorption region of the polystyrene specimen exposed to hydrochloric acid appear to be toward lower frequencies versus the expected higher frequency shift seen in poly (4-chlorostyrene). The absorptions at approximately 706 and 764 cm^{-1} appear at approximately 690-698 cm^{-1} and 748-756 cm^{-1} whereas poly(4-chlorostyrene) exhibits absorptions at approximately 730 and 750 cm^{-1}. The absorptions at 907 cm^{-1}, 1022 cm^{-1}, 1067 cm^{-1}, 1290 cm^{-1}, 1322 cm^{-1}, 1370 cm^{-1}, 1464 cm^{-1} and 1492 cm^{-1} in the polystyrene spectrum (Figure 1) remain essentially unchanged by exposure to hydrochloric acid. Some minor shifts in the absorption spectrum are noticed, however, these are attributed to a reasonable variance in data acquistion. It was therefore concluded that polystryene is essentially resistant to hydrochloric acid at 25°C.

The infrared spectrum of the SMA coplymer specimen exposed to hydrochloric acid is shown as Figure 8. The shifts in absorptions in the range of 665-910 cm^{-1} are again toward lower frequencies as are seen in the spectrum of polystyrene (Figure 7). The peaks at 1447 cm^{-1}, 1496 cm^{-1}, 1601 cm^{-1} and 1780 cm^{-1} (Figure 3) are essentially unchanged by exposure to concentrated hydrochloric acid at 25°C wherein absorptions at 1452 cm^{-1}, 1494 cm^{-1}, 1601 cm^{-1} and 1778 cm^{-1} are observed. Therefore, it was concluded that the SMA did not exhibit a significant increase in reactivity toward hydrochloric acid.

The infrared spectrum of the SMA ester specimen exposed to hydrochloric acid is shown (Figure 9). The absorptions seen in the reference spectrum (Figure 3) appear to be essentially unchanged with the possible exception of the carbonyl absorptions previously reported at 1714, 1747 and 1780 cm^{-1}. In the specimen exposed to hydrochloric acid, absorptions in the carbonyl region appear at 1734, 1784 and 1854 cm^{-1}. The absorptions at 1784

Table IV. Comparison of Infrared Absorptions

Polystyrene cm^{-1} (Figure 1)	Poly (4-chlorostyrene)[2] cm^{-1}	Specimen exposed to HCl cm^{-1}
706	730	690.5, 694.4, 698.0
	750	748.1, 752.2, 756.1
764		
822	828	
844		
907	930	907
1022	1015	1028
1067	1040	1068
1152	1100, 1130	1153
1182	1160, 1190	1180
1290		1302
1322		1322
1370	1365	1366
1464	1450	1450.5
1492	1490	1491.0
1601	1601	1600.9

[1] peak absorptions very strong, may contain multiple absorptions

[2] absortions estimated from standard spectra (Aldrich)

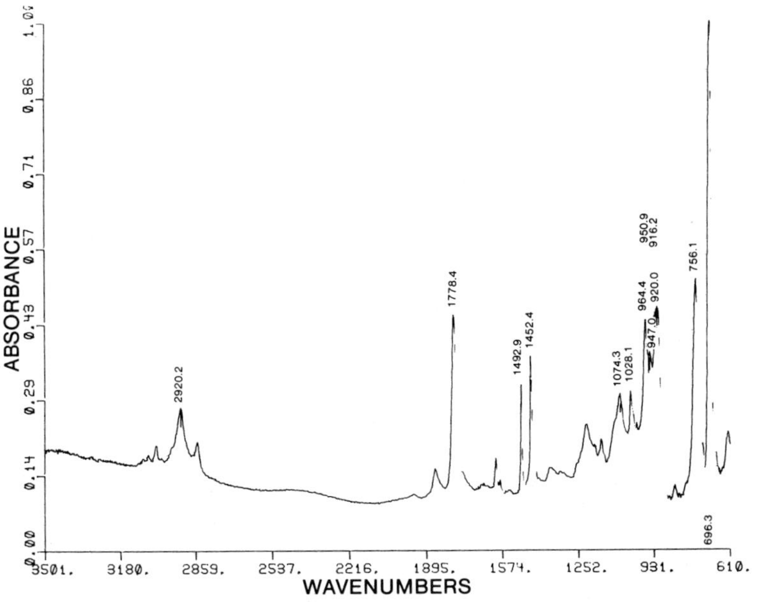

Figure 8. IR spectrum of the styrene–maleic anhydride copolymer exposed to hydrochloric acid.

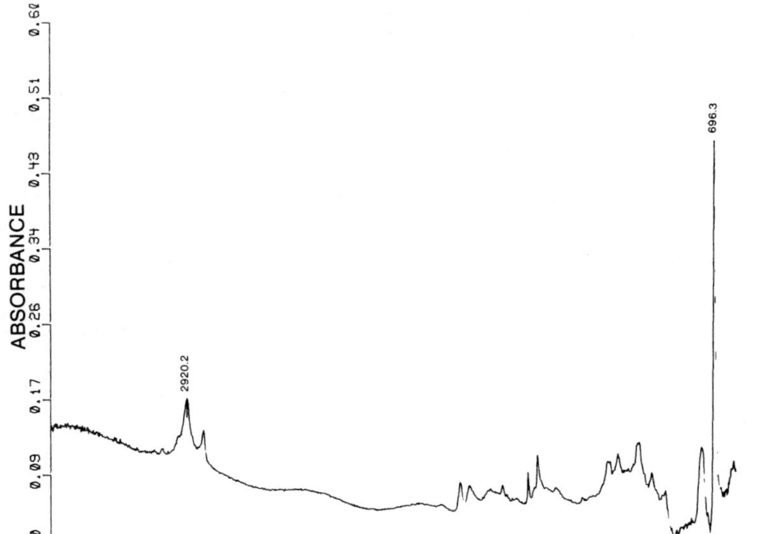

Figure 9. IR spectrum of the styrene–maleic acid/ester exposed to hydrochloric acid.

and 1854 cm^{-1} indicate the presence of a cyclic nonconjugated five-membered anhydride wherein the absorption at the lower frequency, 1784 cm^{-1}, is the stronger peak. Further, as a result of the absence of a clearly indentifiable absorption at 1714 cm^{-1} and 1747 cm^{-1}. These observed shifts were significant enough to conclude that the SMA ester was hydrolyzed when exposed to concentrated hydrochloric acid.

Because the data obtained in the case of exposure to hydrochloric acid at 25°C were difficult to interpret and exhibited very slight polymer/acid interaction, the experiments, wherein the polymers were exposed to glacial acetic acid, were run at 60°C for seven days in an effort to better define the effects of acids on polystyrene and the SMA copolymers under investigation.

The polystyrene specimen exposed to glacial acetic acid was removed from a 60°C oven and the solution was quenched in an effort to slow diffusion of acetic acid out of the sample (Figure 10). The specimen was approximately 25% of its previous size and was severely crazed. The infrared spectrum shows the presence of acetic acid approximately three days after removal and cleaning. The absorption peaks at 1030^{-1}, 1050^{-1}, 1260^{-1}, 1310^{-1}, 1420^{-1} and at approximately 1735 cm^{-1} are indicative of acetic acid diffusing from the test specimen. The previously unseen peak at 1700 cm^{-1} suggests the presence of a conjugated carbonyl wherein the aromatic ester is shifted from a position at 5.75μ (1740 cm^{-1}) seen in aliphatic esters to 5.9μ (1725-1695 cm^{-1}). A very small absorption band is seen at approximately 910 cm^{-1} which is also indicative of an aromatic ester carbonyl. The absorption band at 13μ (770 cm^{-1}) in many of the para and 1,2,4-substituted aromatic esters is also present in the spectrum presented. Therefore it was concluded that glacial acetic acid attached the polylstyrene at a ring position perhaps in the para-position.

The SMA copolymer when exposed to glacial acetic acid was whitened, but displayed a lesser degree of crazing. The infrared spectrum (Figure 11) again shows the presence of acetic acid and presents a much stronger absorption at 910 cm^{-1} and 1860 cm^{-1} are present but their relative ratio when compared to the 1690 cm^{-1} peak are much less. When comparing this spectrum to the spectrum of polystyrene exposed to the same enviroment (Figure 10), it is obvious that the presence of the maleic anhydride group increases the reactivity of the polymer to glacial acetic acid.

The spectrum of the SMA ester exposed to glacial acetic acid is shown as Figure 12. As in the previous two spectra, acetic acid is present after three days, slowly diffusing out of the polymer specimen. The absorption band present at 1840 cm indicates the presence of unreacted (esterified) maleic anhydride as seen in Figure 3. The absorptions at 1714, 1747 and 1780 cm^{-1} are absent suggest that the glacial acetic acid

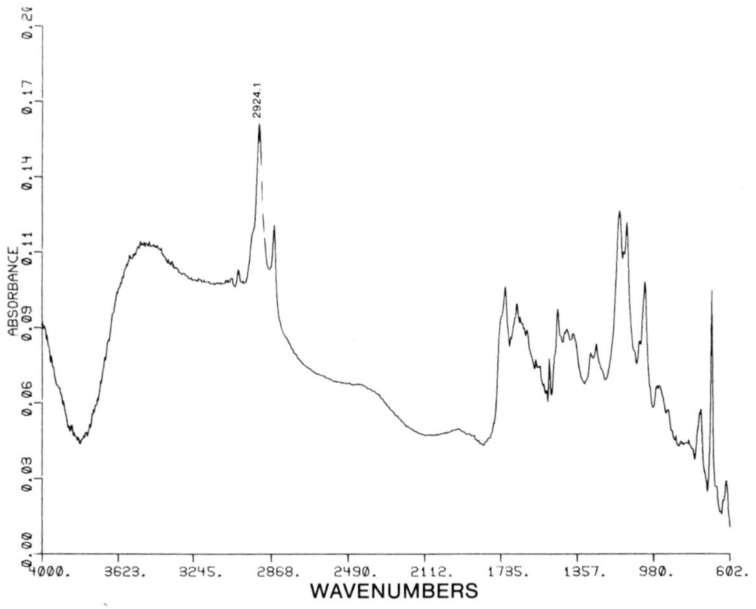

Figure 10. IR spectrum of polystyrene exposed to glacial acetic acid.

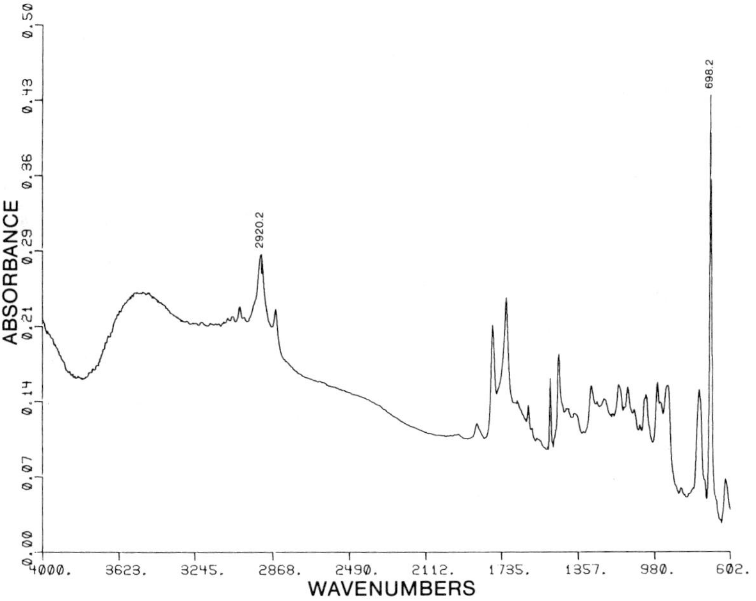

Figure 11. IR spectrum of the styrene–maleic acid anhydride copolymer exposed to glacial acetic acid.

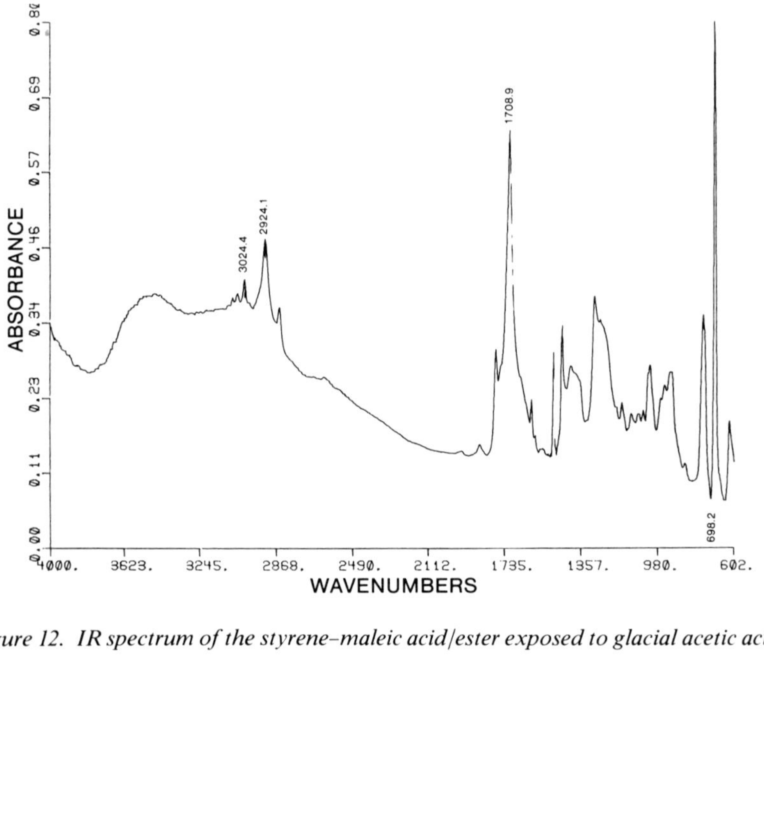

Figure 12. IR spectrum of the styrene–maleic acid/ester exposed to glacial acetic acid.

was strong enough to hydrolyze the ester groups to their respective acid moeities as in the case of hydrochloric acid (Figure 9). This is supported by the very strong absorption at 1708.9 cm^{-1}. The small shoulder at approximately 1700 cm^{-1} indicates that there was some attack on the ring but to a lesser degree than in the case of the SMA copolymer. The acetic acid solution was badly discolored after removal from the oven which suggest the presence of the alcohol previously attached to the polymer or its by-product of reaction.

As seen in Table III, the two SMA copolymers investigated exhibited varying degrees of resistance to the oxygenated solvents to which they were exposed. If one employs the generalized rule cited by Seymour (35) that compatability or solubility on the molecular level can be exected if $\delta_1 - \delta_2$ is less than ± 1.8 H and that the swelling of polymers occurs when $\delta_1 - \delta_2$ is equal to ± 3.2 H (36) then one would not expect solubility of the copolymers under investigation in alcohols with the possible exception of ethanol, 1-propanol and benzyl alcohol wherein $\delta_1 - \delta_2 \approx 1.8$ H. This was the observation with the exception of benzyl alcohol wherein $\delta_{calc.} = 11.4$ H. In both cases, benzyl alcohol completely solubilized the polymers forming homogeneous solutions which were bright and clear. It is interesting to note, however, that methanol and ethanol exhibited a hostile effect on the styrene-maleic anhydride copolymer after several weeks of eposure.

The ketones, as a general class, were expected to solubilize the SMA copolymers. However, acetone, formed a turbid solution which contained a significant quantity of insoluble polymer as a precipitate in the case of the SMA ester. To determine whether the turbidity was the result of concentration, the samples were heated to 55°C which did not result in a uniform polymer/solvent system as seen in the case of benzyl alcohol. Therefore, it was concluded that the ketones reported in Table III exhibited a visibly hostile interacton although the precipitate observed may have been unreacted maleic anhydride or styrene-maleic acid copolymer.

The esters, methyl formate, ethyl acetate and propyl acetate, demonstrated distinct differences in their reaction with the two polymers. Methyl formate, $\delta = 10.2$, exhibited a definite hostile effect on the polymers although $\Delta\delta$max. = -0.73, whereas ethyl and propyl acetate solubilized both polymers. As in the case of the ketones, the latter esters formed turbid solutions which did not clear upon heating however they showed no precipitation upon cooling suggesting the presence of some foreign material.

Perhaps the most interesting of the solvents evaluated were the ethers. Methyl-t-butyl ether exhibited a significant hostile blistering of both test specimens, whereas dioxane and tetrahydrofuran solubilized them. Dimethylformamide and dimethyl sulfoxide both caused the test specimen to disintegrate forming turbid solutions which appeared to be uniform suspensions of polymer in solvent without noticeable precipitation, again suggesting the possible presence of unreacted maleic anhydride or a styrene-maleic acid polymer.

In conclusion, the data presented show that SMA copolymers demonstrate increased reactivity toward nitric acid and essentially equivalent reactivity toward hydrochloric and glacial acetic acid with the exception of the SMA ester which was hydrolyzed.

When considering the generalized rules cited by Seymour (35,36) one would expect molecular compatibility or solubility between the SMA copolymers investigated and the oxygenated solvents reported in Table III with the exception of methanol, 1-butanol, dimethylformamide and dimethyl sulfoxide. If one considers the data reported by Rider (29,30), then an interaction is expected between the two polymers and all the solvents with the exception of the alcohols. However, there is at least one exception to both general theories and this is the interaction of methanol, $\delta_1 - \delta_2 = 4.85$ and HBP - +0.35, with the SMA copolymer.

The reaction of an amorphous polymer with a solvent is dependent upon diffusion of the attacking reactant. If the polymer is at a temperature above its Tg there is a constant cooperative segmental motion that favors diffusion from hole to hole in the polymer mass wherein hole formation in a polymer above its Tg is related to the relative mobility of the diffusing molecules. The rate of diffusion is dependent on the temperature, the molecular size, and the extent of the compatibility of the corrosive molecules with those of the polymer (36).

Although the equal reactivity principle can be used as a guide for the rate of reaction of accessible groups, reactions with solid polymers may be hindered or enchanced by the presence of neighboring group (37). This is the case in the hostile interaction of methanol and ethanol on the SMA copolymer where the anhydride group appears to react with the lower alcohols to form half acid/ester.

Literature Cited

1. Wagner-Jauregg, T., Ber. Dtsch Chem. Ges., 1930, 63, 3213.
2. Mark, H.F., Immergut, B., Immergut, E.H., Young, L.J., Beyon, K.I., "Polymer Handbook", J. Brandup and E.H. Immergut, eds., Interscience Publishers, New York, 1975, Chap. 2, pp. 6-7.
3. Seymour, R.B., Harris, F.H., and Branum, I., Ind. and Eng. Chem., 1949, 41, 509.
4. Seymour, R.B., Owen, D.R., and Kincaid, P.D., Popular Plastsics, 1973, 21.
5. Seymour, R.B., Kincaid, P.D., and Owen, D.R., Adv. Chem. Series, 1973, 129, 230.

6. Seymour, R.B., Tatum, S.D., Boriack, C.J. and Tsang, H.S., Texas J. Sci, 1969, 21, 13.
7. Seymour, R.B., Tsang, H.S., Jones, E.E., Kincaid, P.D., and Patel, A.K., Adv. Chem. Ser., 1971, 99, 418.
8. Muskat, I.E., U.S. Patent 3085994, 1963.
9. Muskat, I.E., U.S. Patent 3178395, 1965.
10. Muskat, I.E., U.S. Patent 3338106, 1968.
11. Muskat, I.E., U.S. Patent 3418292, 1968.
12. Muskat, I.E., U.S. Patent 3415979, 1969.
13. Seymour, R.B., Kincaid, P.D., and Owen, D.R., J. Paint Technol., 1973, 45, 33.
14. Lang, J., Pavelich, W., and Clarey, H., J. Polymer Sci., 1963, 1.
15. Wiley, R., and Parish, D., J. Polymer Sci., 1960, 45, 503.
16. Gaylord, N.G., Stolka, M., Takahaishi, A., and Marti, S.J., Macromol. Sci., 1971, A5, 867.
17. Seymour, R.B., and Garner, D.P., Polymer News, 1978, 5, 209.
18. Seymour, R.B., and Garner, D.P., Polymer, 1976, 17, 21.
19. Isaacson, H.V., and Young, D.W., U.S. Patent 3574790, 1971.
20. Marion, D.L., Hall, R.H., and Rusnack, W.C., U.S. Patent 3528935, 1970.
21. Carrow, D.J., and Verdol, J.A., U.S. Patent 3487125, 1969.
22. Isaacson, H.B., and Young, D.W., U.S. Patent 3415745, 1968.
23. Oppenlander, G.G., Science, 1968, 59, 1311.
24. Physical properties reported were supplied by the manufacturers of the polymers investigated.
25. Foster, R.S., and Volkening, V.B., Corrosion, 1959, 15, 85.
26. Seymour, R.B., Tsang, H.S., and Warren, D., Polym. Eng. Sci., 1967, 7, 55.
27. Jellinek, H.H.G., Flousman, G.F., and Kryman, F.J., J. Appl. Polymer Sci., 1969, 13, 107.
28. Seymour, R.B., "Solubility Parameters of Organic Compounds", CRC Handbook of Chemistry and Physics, 61st Ed., CRC Press, Inc., 1980, pp. 687-690.
29. Rider, P.E., J. Appl. Poly. Sci, 1980, 25, 2975.
30. Rider, P.E., "Hydrogen Bonding Potential, HBP: A Model for Predicting Resin Solubilities in Organic Solvents", submitted for publication (1982).
31. Small, P.A., J. Appl. Chem., 1953, 3, 71.
32. Miller, R.F., "Evaluation of Ethylene Copolymers as Pour Depressants", Macromolecular Solutions, Seymour, R.B., and Stahl, G.A., eds., Pergamon Press, New York, 1982, pp. 84-90.
33. Afremow, L.C., Isakson, K.E., Netzel, D.A., Tessari, D.J., and Vandeberg, J.T., eds., "Infrared Spectroscopy: Its Use in the Coatings Industry", Federation of Societies for Paint Technology, Philadelphia, 1969.

34. Muskat, I.E., U.S. Patent 3046246, 1962.
35. Seymour, R.B., and Carraher, C.E., "Polymer Chemistry", Marcel Dekker Inc., New York, 1981, pp. 61-68.
36. Seymour, R.B., Plastics vs. Corrosives, John Wiley and Sons, New York,, 1982, pp. 25-26.
37. Gaylord, N.G., J. Polymer Sci., 1968, 24, 1.

RECEIVED May 3, 1983

14

Effects of Ethanol/Gasoline and Methyl *t*-Butyl Ether/Gasoline Mixtures on Elastomers

ISMAT A. ABU-ISA

General Motors Research Laboratories, Polymers Department, Warren, MI 48090-9055

The effects of ethanol/gasoline and MTBE/gasoline mixtures on swell and tensile properties of selected automotive elastomers were determined and compared to those of methanol/gasoline mixtures. Two gasolines with aromatic content of 30% and 50% were used in this investigation. Equilibrium swell and tensile measurements were conducted using ASTM standard procedures. The results show that in the case of ethanol/gasoline combinations the elastomers were more severely affected by the mixtures than by the pure components. The ethanol/gasoline mixtures are less severe than the methanol/gasoline mixtures in their effects on most fuel resistant elastomers. In the case of MTBE the effects of mixtures with gasoline were an average of the effects of the pure components. The presence of a higher aromatic content in the gasoline resulted in increased swell and hence increased deterioration of tensile properties of elastomers exposed to the gasoline and its mixtures. Addition of benzene to increase the aromatic content resulted in slightly more detrimental effects on nitrile elastomers than the addition of toluene. The data on all elastomers can be explained in terms of the solubility parameter concept.

For most elastomers the ultimate tensile and elongation values were found to be quantitatively related by simple linear equations to the volume fraction of the rubber swelled in mixtures of gasoline and ethanol or MTBE. They were also found to be related to each other by simple logarithmic relationships.

Reprinted, with permission, from *Rubber Chemistry and Technology, 56(1)* **1983,** 169, American Chemical Society.

Methanol, ethanol, and methy t-butyl ether (MTBE) are blended into some commercial gasolines. It was important to determine the effects of such blends on automotive elastomers especially those used in the fuel system. A summary of the results of our studies [1] and detailed data on the effects of methanol/gasoline mixtures on sixteen elastomers [2] have already been published. This paper presents detailed results of our investigation of the effects of ethanol and MTBE mixtures with gasoline on mechanical and swell properties of automotive elastomers. Comparisons will be made with the effects of methanol/gasoline mixtures on elastomers.

Experimental

The elastomers investigated in this study are shown in Table I along with their percent volume swell in Indolene HO-III.

TABLE I

Elastomers Investigated in Ethanol/Gasoline and MTBE/gasoline Mixtures

	Elastomer	% Swell in Indolene HO-III
1.	Fluorocarbon (*)	0
2.	Polyester Urethane (*)	11
3.	Fluorosilicone	14
4.	Epichlorohydrin Homopolymer (*)	19
5.	Butadiene-Acrylonitrile (*)	34
6.	Polyacrylate	44
7.	Chlorosulfonated Polyethylene (*)	49
8.	Ethylene-Acrylic	84
9.	Ethylene-Propylene-Diene Terpolymer (EPDM) (*)	137
10.	Natural Rubber	169

(*) Elastomers also investigated in MTBE/gasoline mixtures.

According to our previous definition [2] five of the elastomers are considered to be fuel resistant, three medium fuel resistant and two poor fuel resistant. The study of the effects of ethanol/gasoline mixtures was carried out for all ten elastomers. The study of MTBE/gasoline mixtures was conducted for six of the elastomers shown with a star in Table I. The elastomers were formulated using conventional curing agents, oils, carbon black and other ingredients. The formulations of eight of the elastomers have already been published in a previous paper [2]. The formulations of the remaining two elastomers, namely, epichlorohydrin homopolymer and ethylene-acrylic elastomer are shown in Table II.

TABLE II

Formulations of Epichlorohydrin Homopolymer and Ethylene-Acrylic Elastomers

Ingredients	Concentration (phr)	
1. Epichlorohydrin Homopolymer (x21889-33)*		
	1	2
Herclor H	100	100
Stearic Acid	1.0	1.0
Carbon Black N787	90	-
Carbon Black N550	-	40
Dioctyl Phthalate	10	10
Red Lead	5.0	5.0
Irganox 1024**	2.0	2.0
2-Mercaptoimidazoline	1.2	1.2
2. Ethylene-Acrylic Elastomer (91950-296)***		
VAMAC B-124		124
Octadecyl Amine		0.5
Stearic Acid		2.0
Carbon Black N760		35
Vanfre UN****		2.0
Hexamethylene Diamine		1.25
Di-O-tolylguanidine		4.0

* Slabs of these compositions were obtained from Hercules.

** 1,2-bis[(3,5-di-tert-butyl-4-hydroxy) hydrocinnamylamide]hydrazine

*** Slabs of this composition were obtained from DuPont.

**** A fatty alcohol phosphate mold release agent supplied by R. T. Vanderbilt.

As in the methanol invetigation two gasolines were used, namely, Indolene HO-III containing 30% aromatics, and "spiked" Indolene containing 50% aromatics. The second gasoline was prepared by adding 40 ml of toluene to 100 ml of Indolene HO-III. In one experiment 40 ml of benzene were added to 100 ml of Indolene HO-III in order to determine the effects of aromatic type on properties of elastomers exposed to high aromatic gasolines. Mixtures of ethanol with Indolene HO-III and with "spiked" Indolene HO-III were prepared containing 5, 10, 15, 25, 50 and 75% alcohol. Mixtures of similar concentrations were also prepared with MTBE.

Three tensile bars and two volume change specimens of each elastomer were exposed in a test tube to each of the pure fuels and the mixtures. Exposure for 72 hours at room temperature was determined to be sufficient for most elastomers to approach equilibrium swell. Tensile properties were determined using the ASTM D412 procedure for original samples and fuel exposed samples immediately after removal from the fuel. Swell measurements were conducted per the ASTM D471 procedure. To obtain the amount of extractables the fuel swelled samples were dried in a vacuum oven at 100°C for 24 hours and then weighed. In most samples constant weight was achieved within this time period.

The index of refraction was measured at 15°C using Abbe-3L Bausch and Lomb refractomer and a Haake FK constant temperature bath. Refractive index was first measured for ethanol, Indolene HO-III, and mixtures of both components of known concentrations. From this data calibration curves were obtained to relate component concentration with refractive index. These calibration curves were used to determine whether or not preferential absorption of one component of the mixture takes place during elastomer swelling due to permeability considerations, or after equilibrium swell is established due to solubility considerations.

Results and Discussion

Effects of the Composition of Ethanol/Gasoline Mixtures on the Volume Swell and Tensile Properties of Elastomers. The complete results on the effects of various ethanol/Indolene blends on the ten elastomers investigated in this study are shown in Table III. Most of the elastomers exhibit a maximum in the volume change (swell) versus concentration of ethanol plots as seen in Figure 1 for butadiene-acrylonitrile (nitrile) elastomer. The maximum in volume change is accompanied by a minimum in tensile strength and a minimum in elongation. Quantitative relationships between swell and tensile strength and elongation will be discussed later in this paper.

The value of maximum swell and the location on the curve relative to ethanol concentration depend on the type of polymer (Figure 2). The fluorocarbon elastomer exhibits low swell in ethanol, in Indolene, and in the mixtures. For polyurethane and chlorosulfonated polyethylene the polymers swell slightly to moderately in Indolene and ethanol. The maximum swell of these polymers in mixtures of ethanol and Indolene is moderate (40-56%) and occurs in a mixture containing 15% ethanol. For acrylate elastomer the swell in Indolene is moderate (44%) but the swell in

Table III

Effects of Ethanol/Gasoline Mixtures on Properties of Elastomers After Immersion for 72 Hours at Room Temperature

Elastomer	Gasoline	% Ethanol	Tensile Strength MPa	Elongation %	Modulus at 100% Elongation MPA	Durometer Shore A Points	Volume Change %	Extractables %
Fluorocarbon	-	-	18.3	174	7.3	74	-	-
	Indolene	0	15.2	153	9.5	-	0	0
		5	12.7	153	6.3	-	3	0
		10	12.3	153	5.8	-	3	0
		15	12.7	150	7.0	-	4	0
		25	12.3	150	6.2	-	4	0
		50	13.0	158	6.3	-	4	0
		75	12.9	152	6.7	-	3	0
		100	14.6	160	7.1	-	2	0
	Spiked Indolene	0	15.2	162	6.9	-	3	0
		5	13.8	172	5.9	-	5	0
		10	12.6	185	6.0	-	6	0
		15	11.8	153	5.7	-	7	0
		25	12.5	162	5.7	-	7	0
		50	14.5	168	6.3	-	4	0
		75	14.1	163	6.5	-	4	0
Polyester Urethane		-	23.0	477	4.2	70	-	-
	Indolene	0	20.1	413	4.0	-	11	3
		5	12.9	308	3.3	-	29	3
		10	12.4	272	4.0	-	37	3
		15	12.3	267	3.8	-	39	3
		25	12.0	287	3.7	-	40	3
		50	12.9	285	4.0	-	33	3
		75	13.0	315	3.9	-	31	3
		100	17.1	370	3.7	-	19	3
	Spiked Indolene	0	14.5	318	3.5	-	24	3
		5	11.9	262	3.9	-	41	3
		10	10.6	243	3.5	-	51	3
		15	10.2	217	4.0	-	54	3
		25	8.8	205	3.6	-	55	3
		50	10.6	238	3.5	-	49	3
		75	12.3	288	3.3	-	33	3

Table III--Continued on next page

Table III--Continued

Elastomer	Gasoline	% Ethanol	Tensile Strength MPa	Elongation %	Modulus at 100% Elongation MPA	Durometer Shore A Points	Volume Change %	Extractables %
Fluorosilicone		-	7.5	241	2.8	62	-	-
	Indolene	0	6.7	210	2.6	50	14	2
		5	5.1	201	2.2	47	16	2
		10	4.9	190	2.3	46	18	2
		15	4.7	177	2.3	50	18	3
		25	4.1	163	2.3	43	17	2
		50	4.5	183	2.1	45	18	2
		75	4.3	182	2.1	47	13	2
		100	5.6	232	2.1	49	6	1
	Spiked Indolene	0	6.0	178	2.9	48	16	2
		5	4.9	197	2.1	47	19	3
		10	4.3	178	2.1	44	20	3
		15	4.6	185	2.2	45	20	3
		25	4.7	190	2.1	45	19	3
		50	4.7	180	2.4	48	19	3
		75	4.6	182	2.3	47	12	2
Epichlorohydrin Homopolymer		-	16.1	430	3.0	68	-	-
	Indolene	0	12.4	276	3.1	47	19	3
		5	11.3	255	3.0	45	24	5
		10	11.2	238	3.1	42	28	5
		15	9.1	247	3.2	46	28	5
		25	10.8	237	3.2	42	28	5
		50	10.4	228	3.3	47	24	5
		75	12.4	317	3.0	50	18	3
		100	14.2	332	3.1	55	5	2
	Spiked Indolene	0	9.9	223	3.3	45	36	6
		5	8.0	182	3.3	47	48	6
		10	7.5	165	3.6	42	50	6
		15	6.8	163	3.2	43	50	6
		25	7.3	167	3.4	42	47	7
		50	7.9	190	3.4	43	35	6
		75	10.3	233	3.1	43	24	6

Table III--Continued

Elastomer	Gasoline	% Ethanol	Tensile Strength MPa	Elongation %	Modulus at 100% Elongation MPA	Durometer Shore A Points	Volume Change %	Extractables %
Butadiene-Acrylo-		-	16.4	383	4.9	71	-	-
nitrile (Nitrile)	Indolene	0	9.2	208	4.3	55	34	2
		5	7.2	172	3.9	54	46	3
		10	7.3	195	3.5	49	51	3
		15	7.1	175	3.9	51	52	3
		25	8.1	173	4.6	50	48	4
		50	8.1	175	4.5	54	47	4
		75	9.6	230	4.1	55	33	3
		100	13.0	317	4.0	65	8	4
	Spiked Indolene	0	7.5	157	4.7	46	57	4
		5	7.4	152	4.7	47	68	4
		10	6.3	123	4.6	50	76	4
		15	6.8	133	4.8	47	75	5
		25	6.5	132	4.9	51	75	5
		50	7.3	153	4.7	45	63	4
		75	8.6	188	4.5	52	38	4
Polyacrylate		-	10.2	300	4.4	78	-	-
	Indolene	0	3.7	122	3.2	53	44	1
		5	2.3	63	-	50	77	1
		10	2.2	62	-	44	136	1
		15	2.2	60	-	49	130	2
		25	1.9	47	-	50	158	2
		50	1.7	45	-	48	156	2
		75	2.2	67	-	48	138	2
		100	2.5	80	-	52	101	2
	Spiked Indolene	0	1.8	47	-	50	120	1
		5	1.6	40	-	48	165	2
		10	1.6	45	-	48	156	2
		15	1.9	50	-	50	172	2
		25	1.7	47	-	49	182	2
		50	1.6	43	-	46	179	2
		75	1.7	53	-	51	143	2
Chlorosulfonated		-	17.3	222	13.0	87	-	-
Polyethylene	Indolene	0	10.7	148	6.4	-	49	1
		5	8.8	142	5.7	-	54	1
		10	8.1	132	5.6	-	56	1
		15	9.5	148	5.6	-	57	1
		25	9.0	173	4.2	-	48	1
		50	10.5	187	8.8	-	38	0
		75	12.3	202	5.2	-	17	0
		100	-	-	-	-	1	0

Table III--Continued on next page

Table III--Continued

Elastomer	Gasoline	% Ethanol	Tensile Strength MPa	Elongation %	Modulus at 100% Elongation MPA	Durometer Shore A Points	Volume Change %	Extractables %
	Spiked Indolene	0	5.3	135	3.2	-	77	2
		5	5.9	133	4.1	-	77	2
		10	5.6	112	5.0	-	81	2
		15	6.8	82	-	-	76	2
		25	5.3	130	3.6	-	73	2
		50	7.5	155	3.7	-	52	1
		75	13.2	172	7.0	-	24	0
Ethylene-Acrylic		-	16.5	428	3.3	70	-	-
	Indolene	0	4.6	125	4.1	39	84	5
		5	4.3	98	-	36	130	6
		10	3.8	75	-	39	149	6
		15	3.8	65	-	37	174	7
		25	3.8	68	-	37	187	7
		50	4.1	80	-	37	168	8
		75	5.1	113	4.5	36	108	6
		100	7.3	185	3.2	38	55	4
	Spiked Indolene	0	3.8	80	-	39	174	4
		5	3.6	67	-	39	210	4
		10	3.5	55	-	44	221	4
		15	3.4	57	-	42	233	5
		25	3.5	57	-	38	237	5
		50	3.7	65	-	40	195	6
		75	3.8	88	-	38	119	4
Ethylene-Propylene Diene Terpolymer		-	9.1	217	3.6	67	-	-
	Indolene	0	3.3	88	-	45	137	18
		5	3.9	83	-	48	131	18
		10	3.8	80	-	50	124	19
		15	4.2	92	-	45	114	18
		25	4.3	87	-	48	107	17
		50	4.2	93	-	42	94	10
		75	6.1	147	3.7	47	47	4
		100	10.6	237	3.5	68	13	2
	Spiked Indolene	0	3.8	72	-	46	146	18
		5	3.8	77	-	46	133	18
		10	3.8	75	-	48	118	18
		15	4.3	80	-	50	115	18
		25	4.4	93	-	45	100	17
		50	4.6	95	-	40	79	10
		75	6.6	160	3.7	47	43	4

Table III--Continued

Elastomer	Gasoline	% Ethanol	Tensile Strength MPa	Elongation %	Modulus at 100% Elongation MPA	Durometer Shore A Points	Volume Change %	Extractables %
Natural Rubber		-	17.3	449	3.1	64	-	-
	Indolene	0	5.2	137	3.9	45	169	2
		5	4.2	129	3.3	38	186	3
		10	4.2	129	3.3	37	176	3
		15	4.0	118	3.3	39	168	4
		25	5.0	170	3.0	37	157	4
		50	7.9	286	2.8	43	119	3
		75	11.1	360	2.6	50	41	3
		100	14.7	392	3.0	60	2	2
	Spiked Indolene	0	4.7	125	3.8	40	197	2
		5	3.9	119	3.3	37	207	3
		10	4.1	128	3.3	35	200	4
		15	3.8	120	3.2	38	190	4
		25	4.8	158	2.9	36	166	4
		50	6.7	228	2.9	40	109	3
		75	12	357	2.8	47	38	3

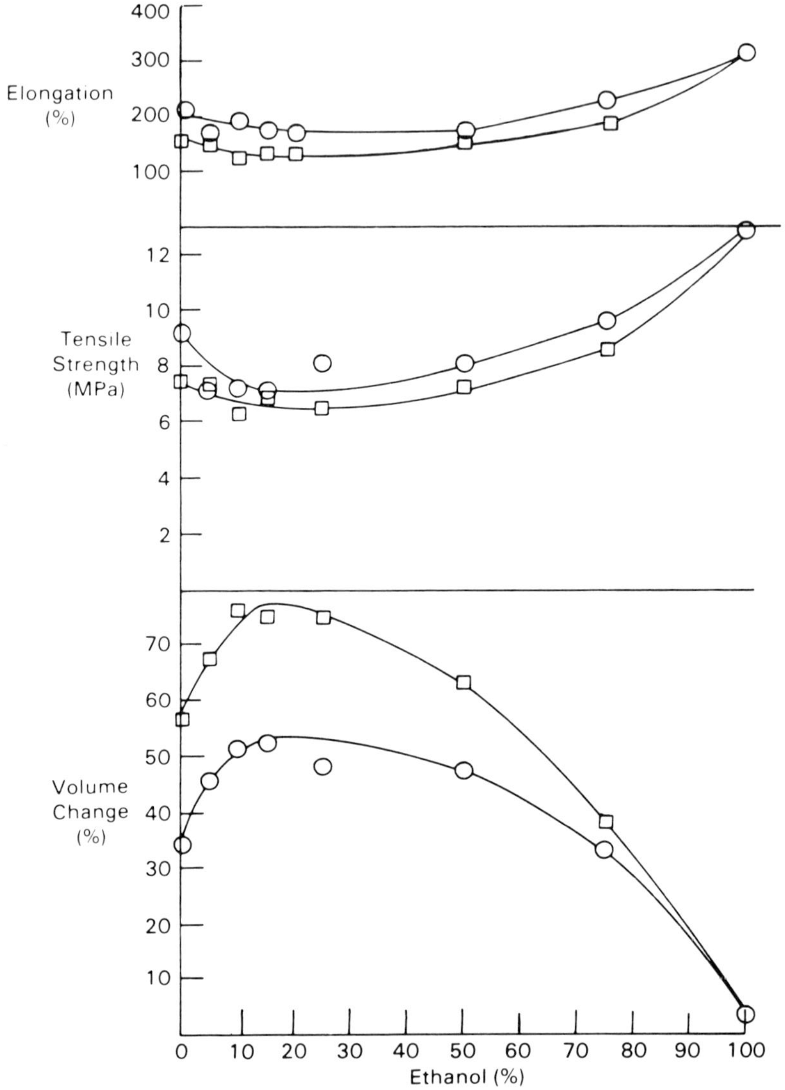

Figure 1. Elongation, tensile strength, and volume change of nitrile elastomer after exposure to ethanol/gasoline compositions. Key: ○, Indolene HO-III; and □, spiked Indolene HO-III.

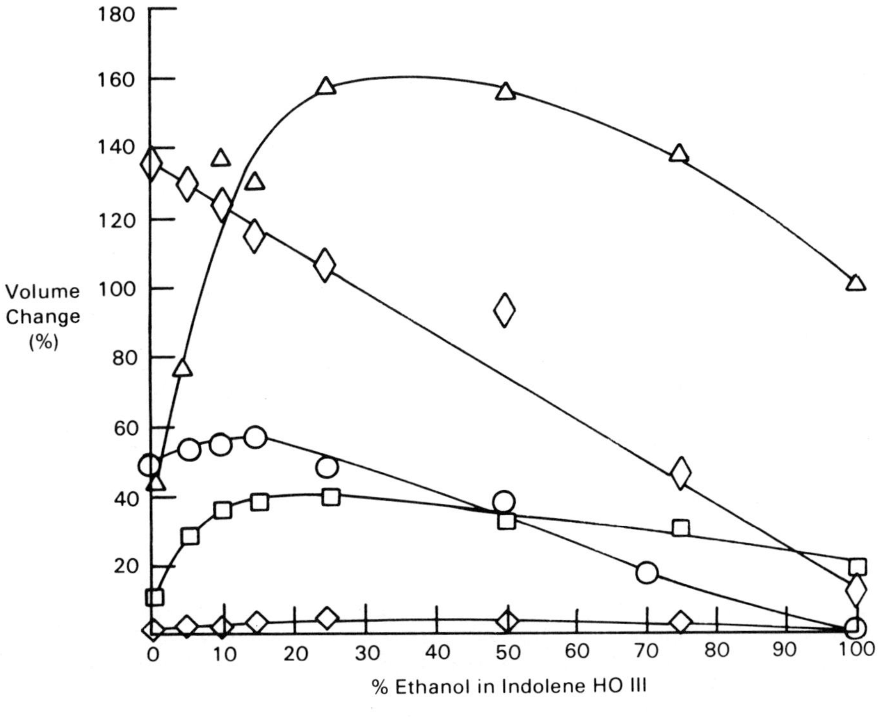

Figure 2. Volume change of elastomers as a function of ethanol concentration in indolene HO-III. Key: ◇, fluorocarbon; □, polyurethane; ○, chlorosulfonated polyethylene; △, acrylate; ethylene–propylene–diene (EPDM).

ethanol is large (100%). The maximum swell for this polymer is high (162%) and occurs in a mixture containing 35% ethanol. For ethylene-propylene-diene (EPDM) elastomer the maximum swell occurs in pure Indolene HO-III and volume swell decreases linearly with increase in ethanol concentration. All the data can be explained in terms of the solubility parameters as will be discussed later in the paper.

Increasing the aromatic content of the fuel mixture results in increased volume swell of the elastomers as seen in Figure 1 and Table IV. Swell values in "spiked" Indolene containing 50% aromatics are consistently higher than in Indolene HO-III containing 30% aromatics. Comparing the volume swell results of spiked Indolene with Indolene HO-III mixtures containing 10% ethanol (columns 4 and 5, Table IV) it is seen that the addition of 10% alcohol is more detrimental than increasing the aromatic content from 30 to 50% in the cases of polyester urethane, fluorosilicone and polyacrylate. The effects are equivalent in the case of fluorocarbon elastomer, whereas in the case of epichlorohydrin homopolymer, nitrile, chlorosulfonated polyethylene, ethylene-acrylic, ethylene-propylene-diene terpolymer (EPDM) and natural rubbers, increasing the aromatic content is more detrimental than adding the alcohol.

For most elastomers the effects of aromatics and alcohol are additive. For example in the case of polyester urethane (Table IV) raising the aromatic content from 30% to 50% increases the swell of the elastomer by 13% (from 11 to 24). The addition of 10% ethanol increases the swell by 26% (from 11 to 37). The addition of both ingredients increases the swell by 40% (from 11 to 51).

For all of the above experiments, the aromatic content was raised to 50% by adding 40 ml of toluene to 100 ml of Indolene HO-III. In one case however we prepared "spiked" Indolene by adding 40 ml of benzene. The effects of benzene spiked Indolene and toluene spiked Indolene on swell properties of nitrile elastomer are shown in Figure 3. The benzene spiked Indolene causes slightly greater volume swell of the nitrile elastomer than the toluene spiked Indolene. Hence the type of aromatic in the gasoline has some effect on the degree of swelling of the elastomers.

Effects of the Composition of Methyl t-Butyl Ether/Gasoline Mixtures on the Volume Swell and Tensile Properties of Elastomers. Detailed results of the effects of various MTBE/gasoline mixtures on the six elastomers are shown in Table V. It is obvious that MTBE/Indolene mixtures affect elastomers in a different manner than the alcohols. The plots of volume swell versus MTBE concentrations do not exhibit maxima. As shown in Figure 4, one elastomer, namely, the fluorocarbon elastomer, shows a sharp increase in swell with increased MTBE concentrations, especially in the concentration range of 25 to 100%. Other elastomers such

TABLE IV

Volume Swell and Tensile Strength of Elastomers After Exposure to Ethanol, Indolene HO-III, Spiked Indolene, and 10% Ethanol Mixtures with Each of the Indolenes

Elastomer	Volume Swell (%) After 72 Hr. Immersion in					Original Tensile Strength MPa	Tensile Strength (MPa) After 72 Hr Immersion in				
				Mixtures of 10% Ethanol in						Mixtures of 10% Ethanol in	
	Ethanol	Indolene	Spiked Indolene	Indolene	Spiked Indolene		Ethanol	Indolene	Spiked Indolene	Indolene	Spiked Indolene
Fluorocarbon	2	0	3	3	6	18.3	14.6	15.2	15.2	12.3	12.6
Polyester Urethane	19	11	24	37	51	23.0	17.1	20.1	14.5	12.4	10.6
Fluorosilicone	6	14	16	18	20	7.5	5.6	6.7	6.0	4.9	4.3
Epichlohydrin Homopolymer	5	19	36	28	50	16.1	14.2	12.4	9.9	11.2	7.5
Butadiene-Acrylonitrile	8	34	57	51	76	16.4	13.0	9.2	7.5	7.3	6.3
Polyacrylate	101	44	120	136	156	10.2	2.5	3.7	1.8	2.2	1.6
Chlorosulfonated Polyethylene	1	49	77	56	81	17.3	-	10.7	5.3	8.1	5.6
Ethylene-Acrylic	55	84	174	149	221	16.5	7.3	4.6	3.8	3.8	3.5
Ethylene-Propylene-Diene Terpolymer	13	137	146	124	118	9.1	10.6	3.3	3.8	3.8	3.8
Natural Rubber	2	169	197	176	200	17.3	14.7	5.2	4.7	4.2	4.1

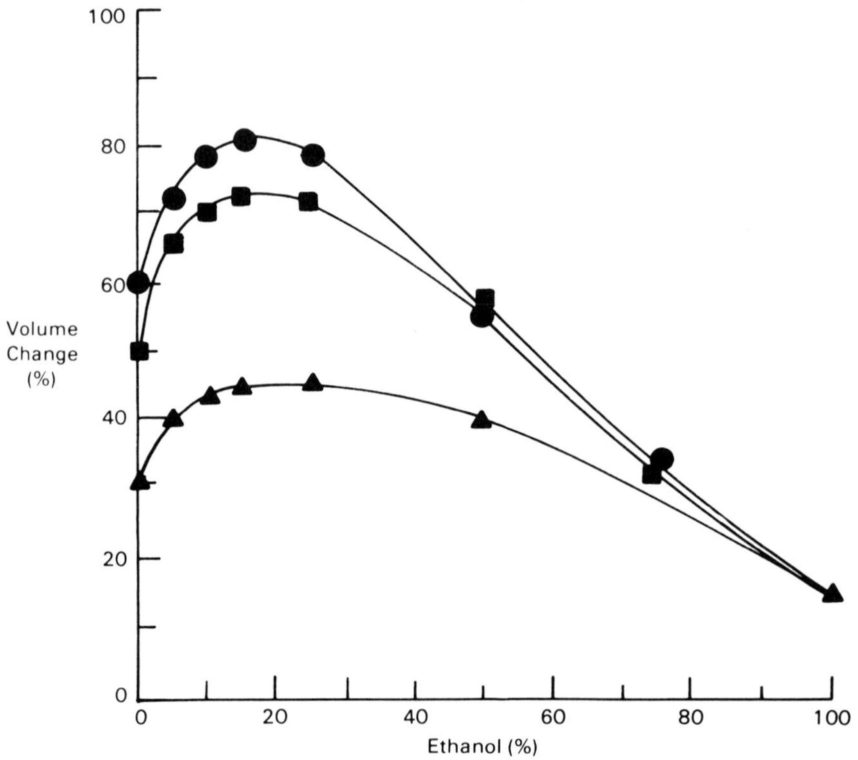

Figure 3. Volume change of nitrile elastomer after exposure to Indolene. Key: ▲, Indolene HO-III; ■, toluene-spiked Indolene HO-III; and ●, benzene-spiked Indolene HO–III.

Table V

Effect of Methyl t-Butyl Ether (MTBE)/Gasoline Mixtures on Properties of Elastomers After Immersion for 72 Hours at Room Temperature

Elastomer	Gasoline	% MTBE	Tensile Strength MPa	Elongation %	Modulus at 100% Elongation MPA	Durometer Shore A Points	Volume Change %	Extractables %
Fluorocarbon		-	18.3	175	7.3	74	-	-
	Indolene	0	14.3	145	7.8	74	1	0
		5	14.9	158	7.2	75	1	0
		10	14.8	160	7.3	76	2	0
		15	14.2	138	7.0	72	3	0
		25	12.0	140	6.8	73	6	0
		50	5.4	82	-	65	34	0
		75	4.4	60	-	60	80	0
		100	4.5	42	-	60	122	0
	Spiked Indolene	0	15.1	152	7.7	74	3	0
		5	13.4	143	7.4	71	3	0
		10	15.2	155	7.3	70	5	0
		15	14.1	147	7.4	71	5	0
		25	10.3	127	7.1	72	10	0
		50	5.0	77	-	64	44	0
		75	3.7	53	-	64	84	0
Polyester Urethane		-	23.0	477	4.2	70	-	-
	Indolene	0	18.3	348	4.8	64	11	1
		5	18.6	362	4.3	64	12	1
		10	17.3	338	4.2	66	13	1
		15	18.9	380	4.1	64	13	1
		25	17.4	340	4.1	60	14	1
		50	15.7	297	4.9	64	17	1
		75	15.6	297	4.2	60	19	2
		100	17.7	340	4.4	63	21	2
	Spiked Indolene	0	15.2	307	4.1	60	23	2
		5	14.9	288	4.4	63	23	2
		10	15.7	318	4.0	60	23	2
		15	17.3	325	4.7	61	24	2
		25	17.3	313	4.3	60	24	2
		50	16.6	313	4.5	62	24	2
		75	17.6	342	4.2	60	23	2

Table V--Continued on next page

Table V--Continued

Elastomer	Gasoline	% MTBE	Tensile Strength MPa	Elongation %	Modulus at 100% Elongation MPA	Durometer Shore A Points	Volume Change %	Extractables %
Epichlorohydrin*		-	12.9	180	9.1	89	-	-
Homopolymer	Indolene	0	10.1	140	8.6	75	14	3
		5	10.4	140	8.5	74	16	3
		10	10.0	125	8.6	74	17	4
		15	10.0	135	8.6	72	17	4
		25	10.5	135	8.5	72	19	4
		50	10.2	113	8.7	73	22	6
		75	10.0	118	8.5	68	31	6
		100	9.1	103	8.6	66	44	7
	Spiked Indolene	0	9.9	117	8.5	70	23	5
		5	10.0	118	8.5	67	24	6
		10	10.1	120	8.6	68	25	6
		15	10.0	121	8.4	67	26	6
		25	10.0	116	8.5	69	27	6
		50	9.7	118	8.3	67	32	6
		75	9.3	109	8.4	66	39	6
Butadiene-Acrylo-		-	14.9	350	4.5	70	-	-
nitrile (Nitrile)	Indolene	0	8.0	222	3.5	56	33	2
		5	7.9	210	3.8	56	33	2
		10	8.0	208	3.8	57	34	2
		15	7.9	198	4.0	57	35	2
		25	7.6	195	3.9	60	37	3
		50	7.5	180	4.1	57	40	3
		75	7.8	185	4.2	58	44	3
		100	7.7	170	4.5	59	46	4
	Spiked Indolene	0	6.4	135	4.7	56	53	3
		5	6.6	152	4.3	53	68	3
		10	6.5	142	4.6	52	67	3
		15	6.6	153	4.2	53	66	3
		25	6.5	150	4.3	54	66	3
		50	7.1	157	4.3	55	65	4
		75	6.7	152	4.0	55	65	4

* Formulation No. 21889-33-1 which is different from the one used for the ethanol/gasoline investigation.

Table V--Continued

Elastomer	Gasoline	% MTBE	Tensile Strength MPa	Elongation %	Modulus at 100% Elongation MPA	Durometer Shore A Points	Volume Change %	Extractables %
Chlorosulfonated		-	17.3	222	13.0	87	-	-
Polyethylene	Indolene	0	13.4	97	-	66	45	0
		5	16.0	100	16.0	67	47	1
		10	17.9	122	17.2	68	48	1
		15	21.7	120	18.8	58	46	1
		25	18.4	100	18.4	63	48	1
		50	19.0	103	17.2	58	53	1
		75	20.3	103	20.1	65	55	1
		100	18.9	97	-	68	57	1
	Spiked Indolene	0	18.8	88	-	69	74	1
		5	16.5	78	-	58	71	1
		10	17.4	75	-	70	68	1
		15	14.1	63	-	61	70	1
		25	18.1	73	-	67	67	2
		50	18.8	88	-	67	70	2
		75	15.4	97	-	58	60	1
Ethylene-Propylene-		-	9.1	217	3.6	67	-	-
Diene Terpolymer	Indolene	0	4.1	80	-	39	138	18
		5	3.5	72	-	37	131	18
		10	3.9	78	-	42	139	18
		15	4.0	83	-	40	132	18
		25	4.1	87	-	41	128	18
		50	4.0	83	-	43	108	18
		75	4.4	100	4.4	47	99	18
		100	5.0	112	4.3	43	79	19
	Spiked Indolene	0	3.7	77	-	41	143	18
		5	4.0	82	-	43	141	18
		10	4.4	88	-	39	139	18
		15	4.2	85	-	40	136	18
		25	4.3	90	-	42	133	19
		50	4.0	90	-	43	116	18
		75	4.2	100	4.2	41	103	18

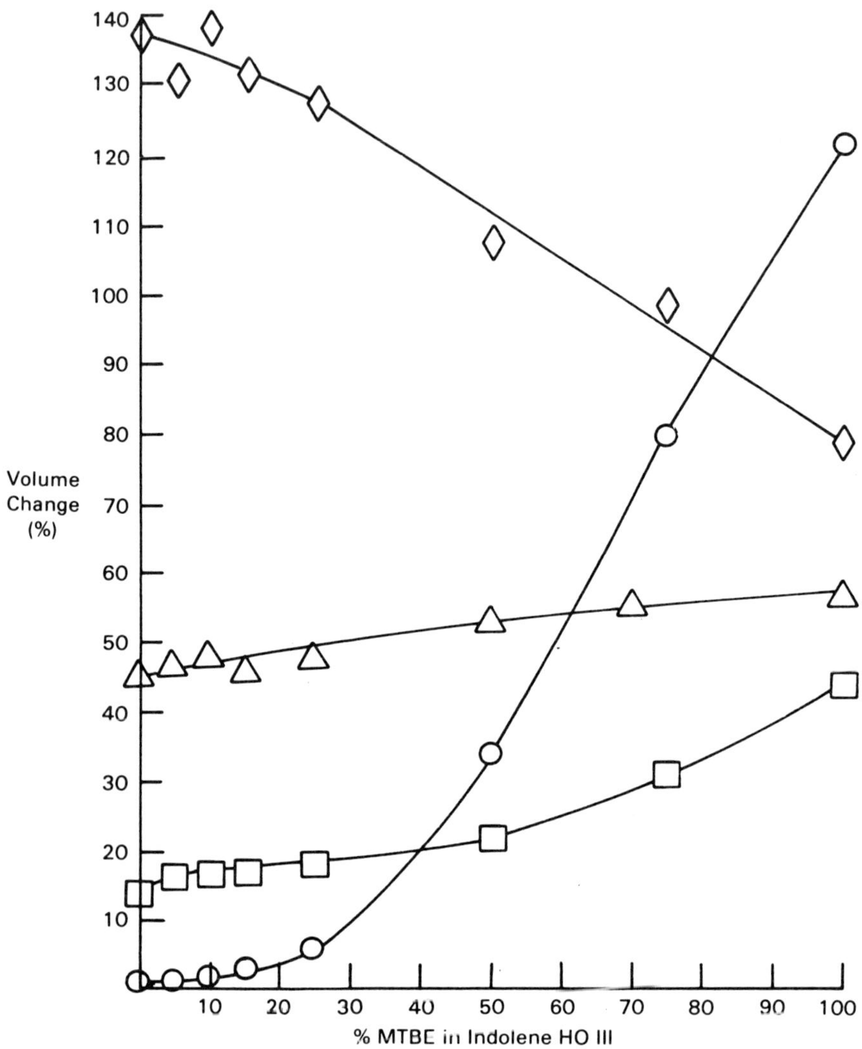

Figure 4. Volume change of elastomers as a function of MTBE concentration in Indolene HO-III. Key: ○, fluorocarbon; □, epichlorohydrin homopolymer; △, chlorosulfonated polyethylene; and ◇, EPDM.

as epichlorohydrin homopolymer exhibit moderate increases in volume swell with increased MTBE concentrations. Some other elastomers exhibit only slight increases in swell, for example, chlorosulfonated polyethylene; still others exhibit a continuous decrease in swell with increased MTBE concentration, for example, EPDM.

Qualitatively MTBE is estimated to have an overall solubility parameter value close to that of Indolene, but has higher polar and hydrogen bonding forces. As a result polar polymers such as fluorocarbon, epichlorohydrin homopolymer and chlorosulfonated polyethylene tend to swell to a greater extent in MTBE rich mixtures, while nonpolar EPDM elastomer swells to a lesser extent in these mixtures. The very large swell of the fluorocarbon in MTBE is not surprising since other ethers such as diethyl ether and dioxane are known to swell the fluorocarbon to a large extent [3].

As in the case of ethanol and methanol a high swell of the elastomers in MTBE mixtures leads to decreased tensile strength and elongation as shown in Table V. Addition of toluene to Indolene HO-III to make "spiked" Indolene leads to increased swell of elastomers in this fuel and its mixtures (Tables V and VI). In the case of all the six polymers investigated increasing the aromatic content from 30 to 50% has a more detrimental effect on elastomers than the addition of 10% MTBE (Table VI).

Comparison of the Effects of Methanol, Ethanol and MTBE Mixtures on Elastomers. The swell behaviors of elastomers in methanol/Indolene mixtures have been adequately explained in terms of the solubility parameter concept [2]. It was found that with the exception of the fluorocarbon elastomer the overall solubility parameter of elastomers was equal to that of the methanol/Indolene mixture in which maximum swell of the elastomer was measured.

Ethanol has an overall solubility parameter [4] of 13.0 $(cal/cc)^{1/2}$. The contributions of the London dispersion (nonpolar) forces, polar forces and hydrogen bonding forces to the overall solubility parameter are δ_d=7.5, δ_p=4.3 and δ_H=10.9 $(cal/cc)^{1/2}$ respectively. The overall solubility parameter of ethanol is slightly lower in value than the solubility parameter of methanol mainly due to smaller contributions by the polar forces. The δ_p for ethanol is 4.3 as compared to 6.0 $(cal/cc)^{1/2}$ for methanol. The swell behavior of most elastomers in ethanol resembles that in methanol. In order to examine whether or not the swell data in ethanol obtained for the various polymers (Table III) can be explained in terms of the solubility parameter concept we compared established solubility parameters of the elastomers with those calculated from maximum swell data. The solubility parameters of nine of the elastomers investigated have been reported in the literature [3-6] and are shown in column 2 of Table VII. The next column of the Table is generated from the present data and lists the concentrations of ethanol in the

TABLE VI

Volume Swell and Tensile Strength of Elastomers After Exposure to Methyl t-Butyl Ether (MTBE), Indolene HO-III, Spiked Indolene HO-III, and 10% MTBE mixtures with Each of the Indolenes

Elastomer	Volume Swell (%) After 72 Hr. Immersion in					Original Tensile Strength MPa	Tensile Strength (MPa) After 72 Hr Immersion in				
				Mixtures of 10% MTBE in						Mixtures of 10% MTBE in	
	MTBE	Indolene	Spiked Indolene	Indolene	Spiked Indolene		MTBE	Indolene	Spiked Indolene	Indolene	Spiked Indolene
Fluorocarbon	122	1	3	2	5	18.3	4.5	14.3	15.1	14.8	15.2
Polyester Urethane	21	11	23	13	23	23.0	17.7	18.3	15.2	17.3	15.7
Epichlorohydrin Homopolymer	44	14	23	17	24	12.9	9.1	10.1	9.9	10.0	10.1
Butadiene-Acrylonitrile	46	33	53	34	67	14.9	7.7	8.0	6.4	8.0	6.5
Chlorosulfonated Polyethylene	57	45	74	48	68	17.3	18.9	13.4	18.8	17.9	17.4
Ethylene-Propylene-Diene Terpolymer	79	138	143	138	138	9.1	5.0	4.1	3.7	3.9	4.4

TABLE VII

Comparision of Solubility Parameters of Elastomers With Solubility Parameters of Ethanol/Gasoline Mixtures at Maximum Swell

Elastomer	Solubility Parameter $(cal/cc)^{1/2}$ of Elastomer from Literature	% Ethanol in Indolene for Maximum Swell	Solubility Parameter $(cal/cc)^{1/2}$ of Ethanol/Indolene Mixture at Maximum Swell
Fluorocarbon	6.6 - 8.7	15	8.7
Polyester Urethane	9.8 - 10.3	25	9.2
Fluorosilicone	8.8	10	8.4
Epichlorohydrin Homopolymer	10.5 - 11.0	15	8.7
Butadiene-Acrylonitrile	9.2 - 9.9	10	8.4
Polyacrylate	9.3	25	9.2
Chlorosulfonated Polyethylene	9.2	10	8.4
Ethylene-Propylene-Diene Terpolymer	7.9 - 8.9	0	7.9
Natural Rubber	7.9 - 8.4	5	8.2

ethanol/Indolene mixtures required for maximum swell of each of the elastomers. Assuming that the solubility parameters of the components in a mixture are additive [2] and using the composition data in column 3, the solubility parameters of the mixtures in which maximum swell occurs are calculated and are listed in the last column of Table VII. They are also plotted versus solubility parameters of elastomers in Figure 5. For all elastomers including the fluorocarbon there is a reasonable agreement with the solubility parameter concept which predicts that at maximum swell the solubility parameter of the elastomer is equal to that of the solvent.

In the case of MTBE the overall solubility parameter is estimated from solubility parameters of similar ethers [3,4,7] to be 7.9 $(cal/cc)^{1/2}$. This is the same value as the calculated solubility parameter of Indolene HO-III [2]. The only difference between the two solvents is the higher contribution of polar and hydrogen bonding forces in the case of the ether. For this reason the degree of swell of most polar elastomers such as fluorocarbon, polyurethane and nitrile is higher in the ether rich mixtures and the degree of swell of nonpolar polymers such as EPDM is higher in Indolene rich mixtures. Because of the similar overall solubility parameter of MTBE and Indolene the plots of volume change versus concentration of MTBE in the mixtures do not exhibit maxima (Figure 4). So unlike ethanol or methanol the effects of MTBE/ Indolene mixtures on elastomers are weighted averages of the effects of the pure components.

A comparison of the effects of mixtures of methanol, ethanol and MTBE on the swell of the fluorocarbon elastomer is shown in Figure 6. It is interesting to note that the effects of methanol are similar to those of MTBE and very different from those of ethanol. This further reinforces the concept of the formation of a hydrogen bonded ether-like structure of methanol [2]. Comparisons of the effects of the alcohols and the ether on other elastomers are shown in Table VIII. For the polar polymers, polyester urethane, fluorosilicone, and butadiene-acrylonitrile, mixtures of Indolene containing 10% methanol produce higher swell than mixtures containing 10% ethanol which in turn produce higher swell than mixtures containing 10% MTBE. For polyacrylate, chlorosulfonated polyethylene, EPDM, and natural rubber, the ethanol mixtures give rise to higher swell values than the methanol mixtures. Similarly the MTBE mixtures give rise to higher swell than methanol mixtures in the case of EPDM and chlorosulfonated polyethylene. All data is in agreement with the predictions of the solubility parameter concept.

Index of Refraction Measurements. The use of index of refraction measurements to study variations in concentrations of alcohol/ gasoline mixtures, during swelling or after equilibrium swell, was attempted. For calibration, the indices of refraction of several ethanol/Indolene HO-III and methanol/Indolene HO-III mixtures were

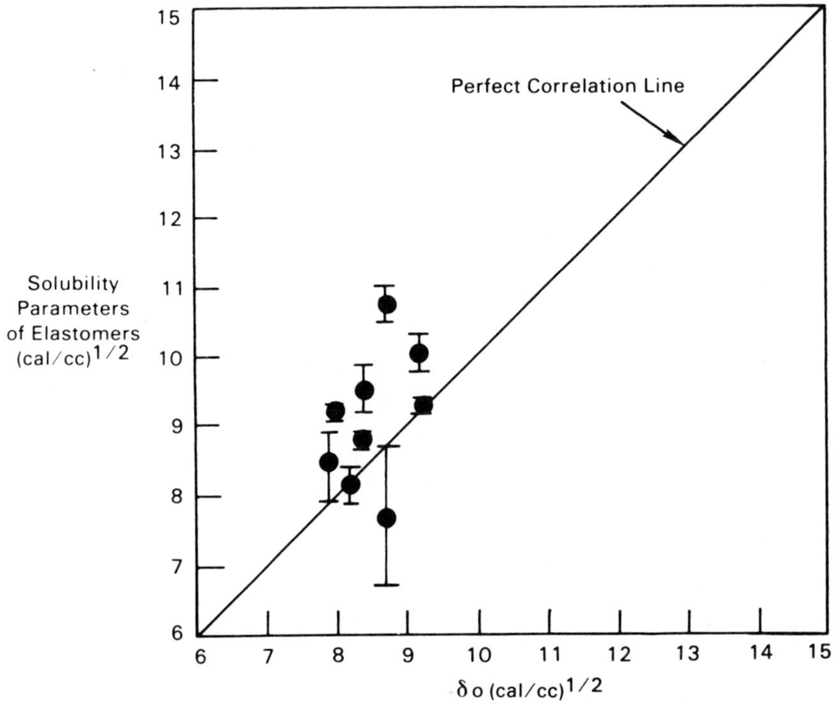

Figure 5. Solubility parameters of elastomers compared to solubility parameters of ethanol/Indolene at maximum swell.

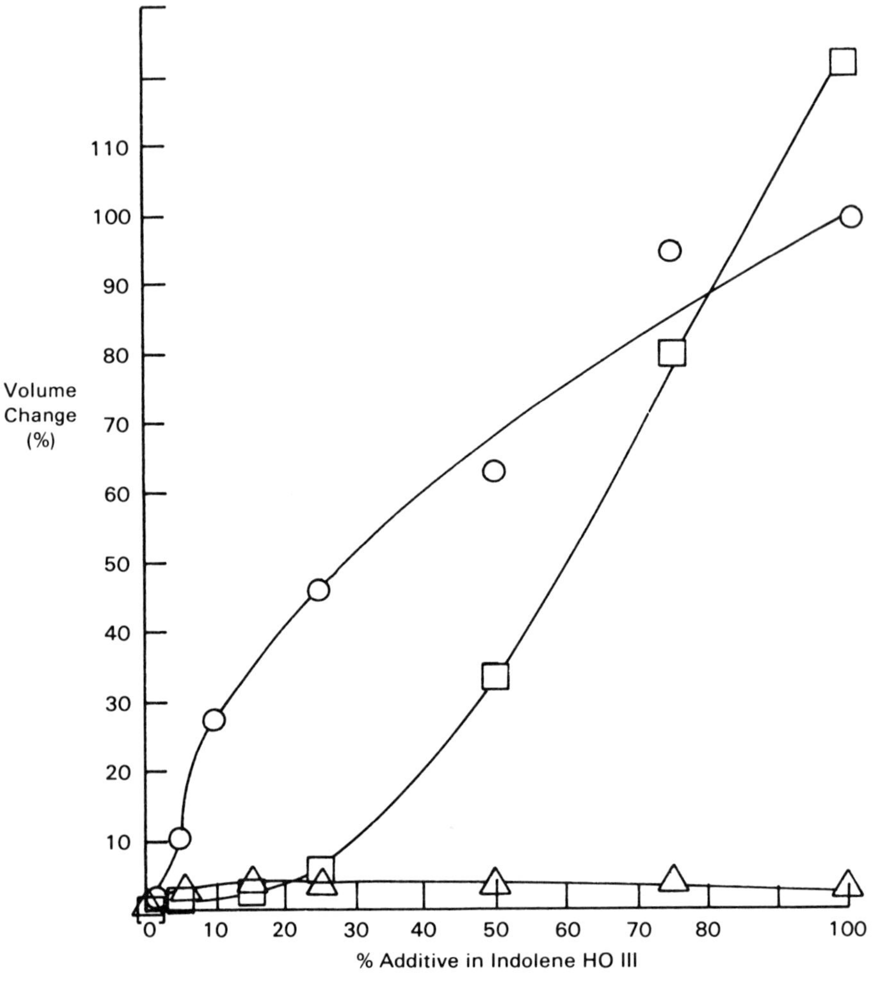

Figure 6. Volume change of fluorocarbon elastomer as a function of additive concentrations in Indolene HO-III. Key: ○, methanol; △, ethanol; and □, MTBE.

TABLE VIII

Comparisons of the Swell of Various Elastomers in Methanol, Ethanol and MTBE Mixtures with Indolene HO-III

Elastomer	Volume Swell (%) After 72 Hr. Immersion In						
					Mixtures Containing 90% Indolene and 10% of		
	Indolene	Methanol	Ethanol	MTBE	Methanol	Ethanol	MTBE
Fluorocarbon	0	100	2	122	27	3	2
Polyester Urethane	11	18	19	21	42	37	13
Fluorosilicone	14	8	6	-	21	18	-
Butadiene-Acrylonitrile	34	14	8	46	53	51	34
Polyacrylate	44	94	101	-	112	136	-
Chlorosulfonated Polyethylene	49	1	1	57	41	56	48
Ethylene-Propylene-Diene Terpolymer	137	0	13	79	109	124	139
Natural Rubber	169	1	2	-	148	176	-

measured. When the indices of refraction were plotted against the alcohol content linear plots were obtained as shown in Figure 7.

In one experiment 17.7 grams of nitrile rubber were immersed in 200 ml Indolene HO-III for 72 hours. The index of refraction of the Indolene was 1.4294 before immersion and 1.4298 after immersion. This means that the immersion of rubber in the Indolene did not significantly change the index of refraction.

Next we wanted to determine if the relative concentrations of the alcohol/gasoline mixtures change after equilibrium swelling of an elastomer in the mixtures due to preferential absorption of one of the components by the elastomer. For this experiment we pipetted into test tubes 50, 100, 150 and 200 ml of a mixture composed of 20% ethanol and 80% Indolene HO-III. We then placed in each of the tubes 2.43 grams of nitrile rubber for 72 hours. After that the indices of refraction of the mixtures were measured. Indices of refraction are plotted versus volume of the mixtures in Figure 8. From the figure it is seen that the indices of refraction of the mixtures have higher values, i.e. the mixtures contain higher concentrations of Indolene, after exposure to the rubber. Hence the nitrile rubber had absorbed a mixture containing a higher concentration of alcohol than that in the original mixture to which it was exposed.

Another experiment was conducted using 2.53 grams of acrylate rubber and immersing them in 200 ml mixtures containing 25% methanol and 75% Indolene HO-III for 0.08, 0.25, 0.5, 1, 5, 24 and 72 hours. As seen in Figure 9 the change in methanol concentration due to preferential absorption in the acrylate rubber is the same for short as well as long exposures.

Both of these studies demonstrate that measurements of refractive indices are useful for studying changes in concentration of alcohol/Indolene mixtures, even after they have been in contact with elastomer samples.

Effect of the Degree of Swelling on Ultimate Properties of Elastomers. In the investigation of the effects of methanol/gasoline blends on elastomers [2] we established that the ultimate stress σ_b and the ultimate elongation E_b in swelled elastomers are linearly related to the volume of rubber v_r in the swelled network, i.e.,

$$\sigma_b = a + bv_r$$

$$E_b = c + dv_r$$

We also found out that logarithmic relationships between σ_b and v_r and E_b and v_r are linear, i.e.,

$$\ln\sigma_b = a' + b' \ln v_r$$

$$\ln E_b = c' + d' \ln v_r$$

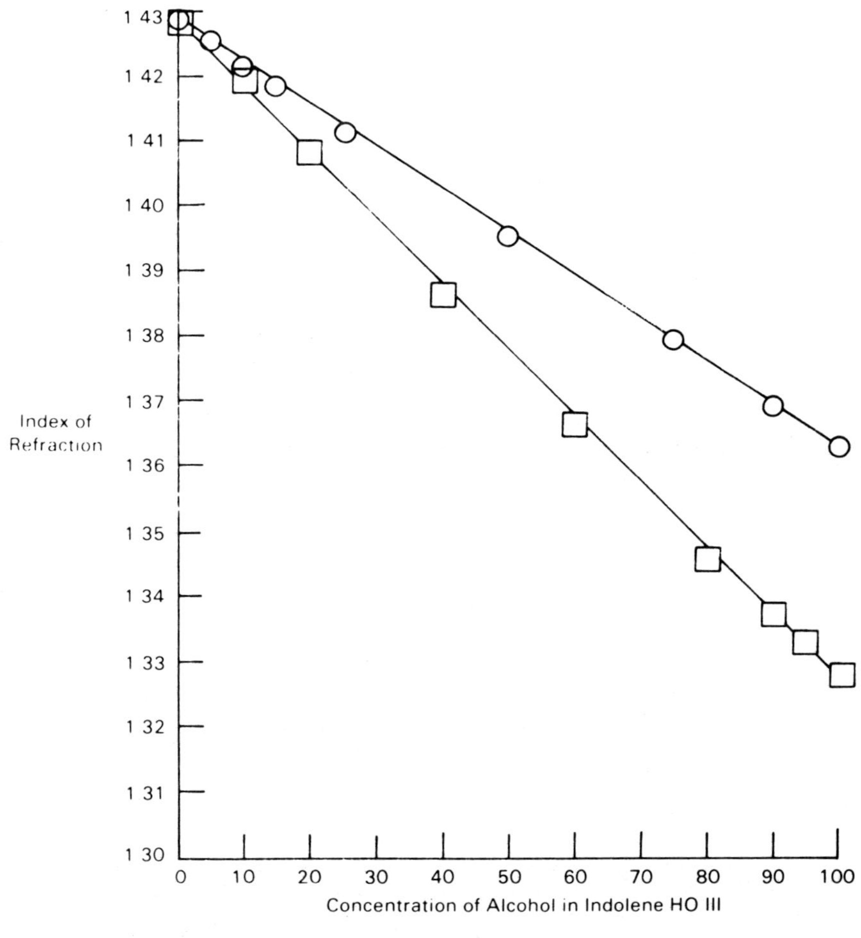

Figure 7. Indices of refraction of alcohol mixtures with Indolene HO-III measured at 15°C. Key: ○, ethanol; and □, methanol.

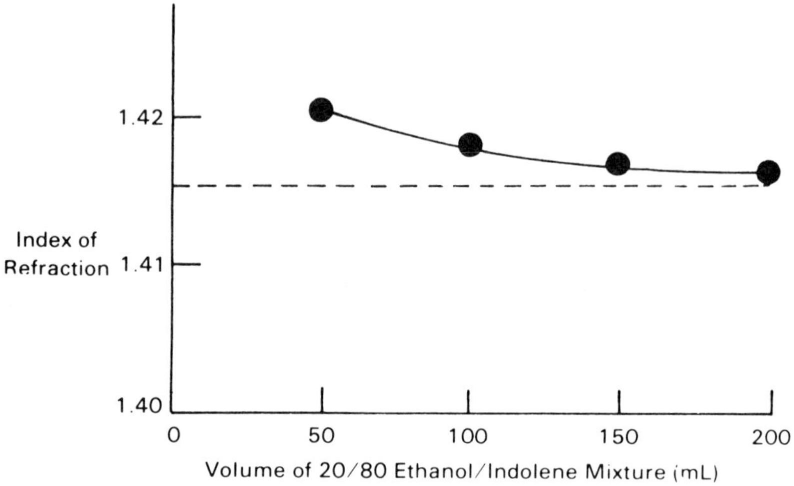

Figure 8. Refractive indices of various volumes of 20/80 ethanol/Indolene mixture before exposure (---) and after exposure (——) to nitrile rubber.

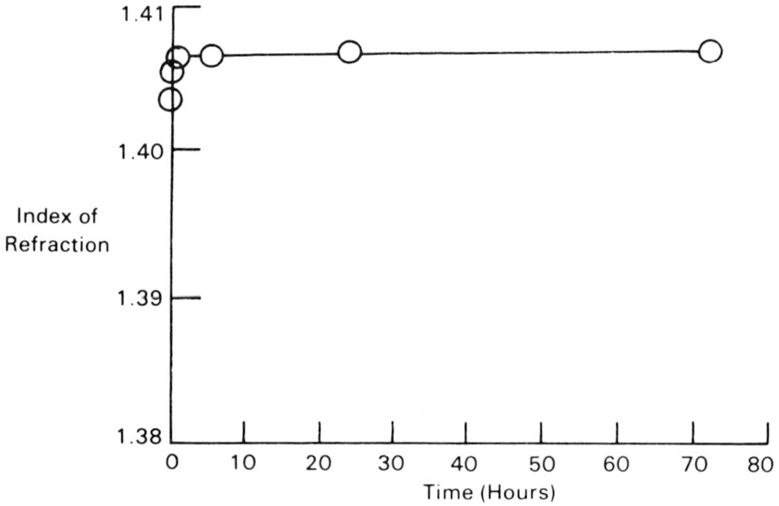

Figure 9. Refractive indices of 25/75 methanol/Indolene mixture before and after exposure to acrylate rubber for various time periods.

Plots of σ_b and E_b versus v_r for EPDM and polyurethane elastomers swelled in ethanol and MTBE are shown in Figure 10. Linear regression analyses for σ_b versus v_r and E_b versus v_r for all elastomers investigated are shown in Table IX for ethanol swelled elastomers and Table X for MTBE swelled elastomers. From the above figure and tables it is evident that linear relationships between σ_b and v_r and E_b and v_r represent the data well for most of the elastomers studied. There is more scatter in the data obtained for the MTBE swelled elastomers because the measured differences in the swell values in mixtures containing varying amounts of MTBE are small and hence the relative error in the swell measurement becomes very large. Hence for some of the elastomers investigated (for example, polyurethane and EPDM) slight differences in the slopes and intercepts of the lines observed in Figure 10 for the MTBE and ethanol swelled polymers are most likely the result of the scatter in the data rather than a true difference caused by the difference in the nature of the swelling agent.

Logarithmic correlations of σ_b versus v_r and E_b versus v_r were also examined as seen in Figure 11 for polyurethane and EPDM elastomers swelled in ethanol and MTBE and in Tables IX and X for all the polymers. Again, the data demonstrate perfect linear relationships.

<u>Relationships Between Tensile Strength and Ultimate Elongation.</u> The actual stress on a rubber sample during tensile testing is defined as $\sigma\lambda$, where σ is the engineering stress and λ is the stretch ratio $\frac{L}{L_o}$. L and L_o are the lengths of the stretched and unstretched samples respectively.

Relationships between actual ultimate tensile strength $\sigma_b\lambda_b$ and the extension ratio ε were examined for the elastomers swelled in ethanol and MTBE. ε_b is defined as $(L-L_o)/L_o$. Plots of $\ln \sigma_b\lambda_b$ versus $\ln\varepsilon_b$ for EPDM and polyurethane elastomers swelled in ethanol and MTBE are shown in Figure 12. It is obvious from this figure that the functions are linearly related for these polymers. Regression linear analyses of the data for all polymers are shown in Table XI. From the R^2 values one can conclude that $\ln\sigma_b\lambda_b$ is linearly related to $\ln\varepsilon_b$ for most polymers. For any one polymer the differences in the slopes and intercepts of the lines for the polymer swelled in ethanol and in MTBE are more likely the result of scatter in the data than real differences caused by differences in the chemical structure of the solvents.

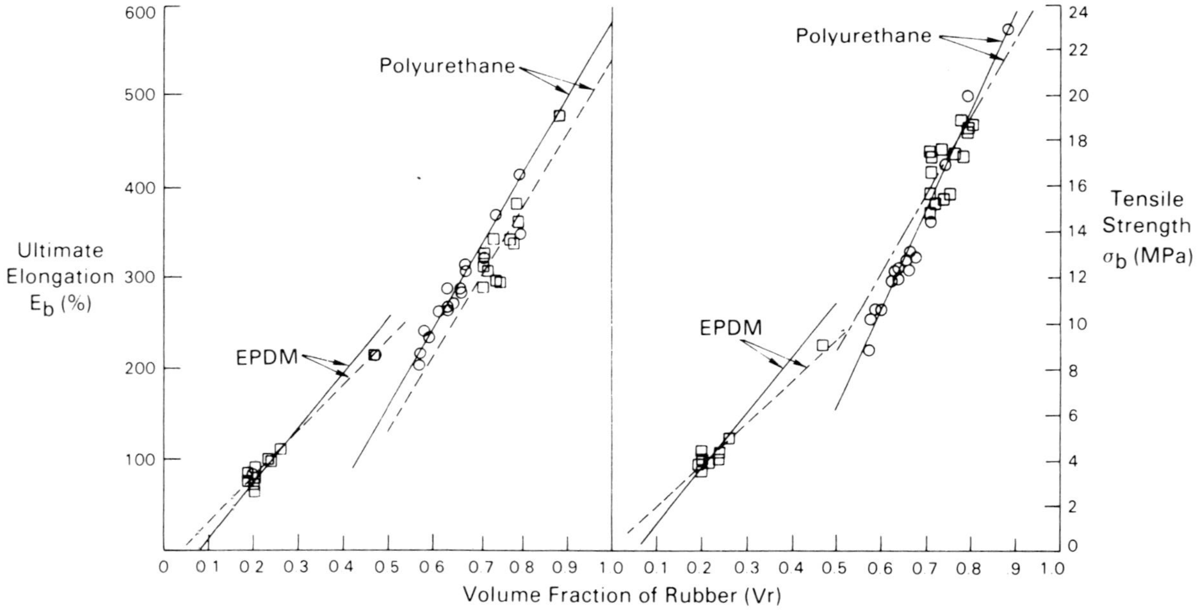

Figure 10. Ultimate elongation and tensile strength vs. volume fraction of polyurethane and EPDM swelled in ethanol (○) and MTBE (□). For the EPDM curve, individual data points were plotted only for MTBE because of overlapping.

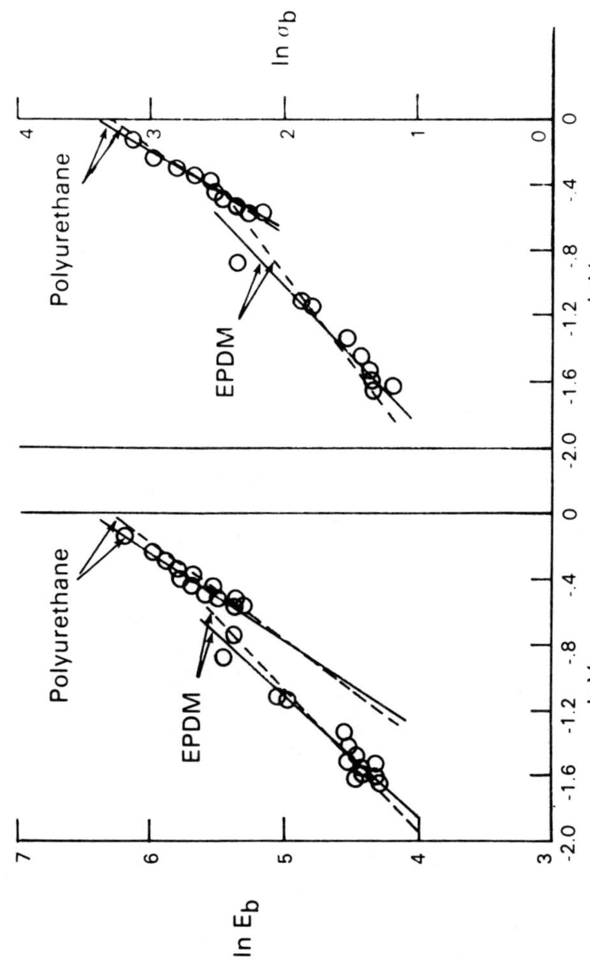

Figure 11. Natural logarithm of ultimate elongation (ln E_b) and tensile strength (ln σ_b) vs. natural logarithm of volume fraction (ln V_r) of EPDM and polyurethane elastomers swelled in ethanol (——) and MTBE (---). Individual data points for MTBE are not shown because of overlapping.

TABLE IX

Linear Regression Analysis of Tensile Strength (σ_b) and Ultimate Elongation (E_b) Versus Volume of Elastomer in Ethanol Swelled Network

Elastomer	σ_b vs. v_r			E_b vs. v_r			$\ell n\ \sigma_b$ vs. $\ell n\ v_r$			$\ell n\ E_b$ vs. $\ell n\ v_r$		
	Slope	Intercept	R^2	Slope	Intercept	R^2	Slope	Intercept	R^2	Slope	Intercept	R^2
Fluorocarbon	88.4	-47.3	0.51	-130	250	0.0029	4.17	4.16	0.51	-0.54	4.88	0.028
Polyester Urethane	43.6	-15.7	0.96	845	-264	0.98	2.03	3.41	0.96	1.84	6.44	0.97
Fluorosilicone	29.3	-7.87	0.52	732	-131	0.66	2.33	3.52	0.48	1.61	6.57	0.61
Epichlorohydrin Homopolymer	34.8	-10.3	0.93	944	-314	0.92	1.99	3.37	0.92	2.20	6.64	0.94
Butadiene-Acrylonitrile	27.6	-5.27	0.92	738	-177	0.96	1.46	3.14	0.93	1.78	6.47	0.96
Polyacrylate	1.72	-2.76	0.90	525	-88.6	0.92	1.55	2.69	0.90	1.69	6.19	0.90
Chlorosulfonated Polyethylene	52.0	-8.42	0.95	487	-12.6	0.83	1.90	4.24	0.91	1.16	6.27	0.75
Ethylene-Acrylic (Vamac)	23.5	-2.62	0.90	689	-113	0.95	1.16	2.87	0.87	1.62	6.43	0.97
Ethylene-Propylene Diene Terpolymer	24.4	-1.26	0.92	610	47.0	0.94	1.20	3.22	0.94	1.35	6.51	0.95
Natural Rubber	25.4	-2.08	0.97	662	-26.4	0.91	1.30	3.25	0.96	1.22	6.57	0.91

TABLE X

Linear Regression Analysis of Tensile Strength (σ_b) and Ultimate Elongation (E_b) Versus Volume of Elastomer in MTBE Swelled Network

Elastomer	σ_b vs. v_r			E_b vs. v_r			$\ell n\ \sigma_b$ vs. $\ell n\ v_r$			$\ell n\ E_b$ vs. $\ell n\ v_r$		
	Slope	Intercept	R^2	Slope	Intercept	R^2	Slope	Intercept	R^2	Slope	Intercept	R^2
Fluorocarbon	34.0	-9.40	0.86	316	- 70.1	0.95	2.00	3.35	0.87	1.73	5.63	0.98
Polyester Urethane	34.0	-8.19	0.67	817	-276	0.72	1.39	3.25	0.62	1.67	6.29	0.68
Epichlorohydrin Homopolymer	17.4	1.79	0.79	402	- 67.1	0.86	0.77	2.88	0.78	1.42	5.87	0.85
Butadiene-Acrylonitrile	24.6	-4.51	0.84	690	-156	0.88	1.32	2.95	0.85	1.68	6.39	0.86
Chlorosulfonated Polyethylene	-0.20	17.7	2.2×10^{-5}	693	-124	0.93	4.77	2.87	2.5×10^{-5}	1.97	6.81	0.85
Ethylene-Propylene-Diene Terpolymer	18.8	0.18	0.95	503	- 19.2	0.97	0.93	2.78	0.90	1.15	6.25	0.93

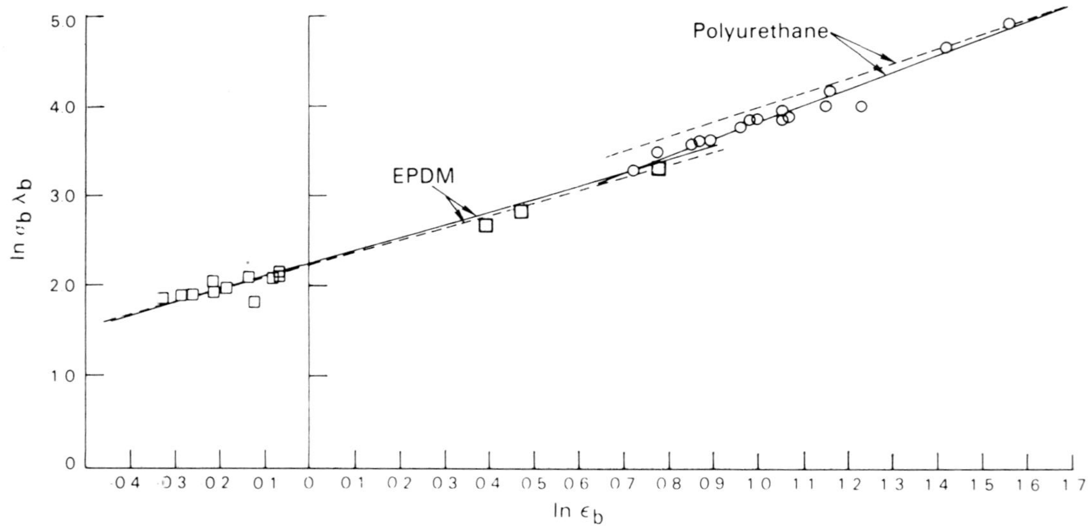

Figure 12. Natural logarithm of ultimate actual stress ($\ln \sigma_b\lambda_b$) vs. natural logarithm of ultimate stretch ratio ($\ln \epsilon_b$) for polyurethane and EPDM swelled in ethanol (——) and MTBE (---). Individual data points for MTBE are not shown because of overlapping.

TABLE XI

Linear Regression Analysis of Tensile Strength Versus Ultimate Elongation for Elastomers Swelled in Ethanol and MTBE

Elastomer	$\ell n\ \sigma_b\ \lambda_b$ vs. $\ell n\ \varepsilon_b$ (Ethanol)			$\ell n\ \sigma_b\ \lambda_b$ vs. $\ell n\ \varepsilon_b$ (MTBE)		
	Slope	Intercept	R^2	Slope	Intercept	R^2
Fluorocarbon	1.41	2.90	0.42	1.66	2.88	0.96
Polyester Urethane	1.84	1.96	0.99	1.62	2.36	0.98
Fluorosilicone	1.94	1.42	0.78	-	-	-
Epichlorohydrin Homopolymer	1.60	2.15	0.98	1.08	2.89	0.94
Butadiene-Acrylonitrile	1.48	2.26	0.97	1.39	2.24	0.93
Polyacrylate	1.41	1.97	0.98	-	-	-
Chlorosulfonated Polyethylene	1.78	2.38	0.78	0.65	3.57	0.71
Ethylene-Acrylic (Vamac)	1.31	2.30	0.96	-	-	-
Ethylene-Propylene-Diene Terpolymer	1.42	2.24	0.97	1.35	2.24	0.97
Natural Rubber	1.70	1.83	0.98	-	-	-

Acknowledgments

It is a pleasure to acknowledge the capable technical assistance of Ms. Dorthie J. McIntyre and Mr. Darrel S. Nelson.

Literature Cited

1. Abu-Isa, I. A.; SAE Technical Paper Series No. 800786, Passenger Car Meeting, Dearborn, MI, June 9-13, 1980.
2. Abu-Isa, I. A.; in "Macromolecular Solutions, Solvent-Property Relationships in Polymers," Seymour, R. B. and Stahl, G. A., Eds., Pergamon Press, New York, 1983; pp. 166-197.
3. Beerbower, A. and Dickey, J. R., ASLE Trans. (1969) 12, 1.
4. Sheehan, C. J. and Bisio, A. L. Rubber Chem. and Techn., 1966, 39(1), 149.
5. Barton, A. F. M. Chem. Reviews, 1975, 75(6), 731.
6. Abu-Isa, I. A. J. Radiation Curing 1981, 8(2), 19.
7. Hansen C. M. and Skaarup, K. J. Paint Techn., 1967, 39, 511.

RECEIVED March 3, 1983

15

Effects of Ozone Exposure on the Physical Properties of Butadiene and Styrene/Butadiene Copolymers

K. E. STEPHENS and C. L. BEATTY

University of Florida, Department of Materials Science and Engineering, Gainesville, FL 32611

Polystyrene and polybutadiene homopolymers as well as random and block copolymers of these mers have been studied via dielectric relaxation spectroscopy and tensile stress-strain measurements. The results suggest that some block copolymer systems studied have styrene rich surfaces which appear to partially crosslink upon initial exposure to ozone even though surface oxygen concentrations are not significantly affected. After continued exposure these samples appear to then undergo chain scission. Complex plane analysis implies that after degradation has occured, the Tg process for PS is reduced in temperature and occurs by the same mechanism as in non degraded PS. The butadiene rich materials appear to undergo predominantly chain scission with increases in oxygen concentrations at the surface of the sample.

Photo-oxidation of polystyrene, PS, (1-5) and high impact polystyrene, HIPS, (6) have been well studied by noting changes in the infrared spectra (1,2,3,6) contact angle (1) ultraviolet spectra (2), dynamic mechanical (6) and dielectric relaxation (4,5) properties. The effect of ozone exposure on the properties of elastomer including styrenebutadiene random (SBR) copolymer has been studied via infrared spectroscopy (7). In addition, the XPS spectra of ozone treated polystyrene, polybutadiene (PB) and block and random copolymers has been reported (8). Ozone effects on relative surface atomic concentrations and surface energy has also been examined and reported for PS, PB and both block and random styrene copolymers (20). In this study we wish to systematically investigate the effects of chemical composition, chain microstructure and morphology with regard to the effect of treatment as manifested by changes in a variety of physical properties.

Experimental

Chloroform solutions prepared using the "as received" polymers

0097-6156/83/0229-0261$06.00/0
© 1983 American Chemical Society

(see Table I) were reprecipitated in cold methanol and vacuum dried overnight to remove residual methanol. Polystyrene, 310 and 308 reprecipitated polymers were redissolved into chloroform and cast using a micrometer adjustable film caster onto glass plates. The index cards had the center portion removed so the index card "picture frame" served as a support for the elastomer films and facilitated handling of the films. The Solprene 308 and 1205 films were removed from the glass plates under water. The films were then placed in a vacuum for at least 3 days. To acquire uncross-linked polybutadiene and random copolymer films of sufficient strength to be handled and cut into specimens, these solutions were poured into a rectangular form atop a glass plate and allowed to dry. Repeated castings were made until sufficient thickness had been built up for the film to retain its own shape upon removal from the glass plate. These films were dried in vacuum for one week.

Polybutadiene crosslinking was done by adding benzoyl peroxide equivalent to 1% of polymer weight in the reprecipitated solution. The mixture was then turned at approximately 30 rpm's for 1 hr. on roller to facilitate mixing, prior to pouring into the form.

Samples were placed into a bell jar having an air flow of 1.5 liters/min. which contained 375 ppm$\pm$10% ozone. For the reciprocity study, ozone concentration was varied from 100 to 1000 ppm $\pm$ 10%. Ozone generation was accomplished using a P.C.I. Model C2P-3C high voltage ozone generator. Ozone concentrations were monitored by bubbling the gas flow through a buffered (6.7 pH) potassium iodide salt solution and titrating with sodium thiosulfate (9). All films were kept in vacuum for at least one additional day.

Upon opening the ozone chamber to introduce samples, a concentration gradient was produced. For this reason the subsequent time involved for equilibrium to be reestablished, exposure time was not used as the only gauge of exposure in reciprocity measurements. Under these circumstances, the area under the concentration vs time curve appears to be a better measure of exposure of the sample to ozone. This area factor was kept constant over a range of concentration conditions.

Tensile Stress-Strain

Tensile testing was done using an Instron Model 1122 load frame and an Instron 1000 lb. maximum combination tension/compression load cell. Samples were cut from films using ASTM 1822 Type D standard cutter for tensile samples. Crosshead speed was 5 inches per minute for all samples. Prior to tensile testing, polystyrene homopolymer tensile samples were annealled at 85°C for 4 days in order to remove any additional internal stresses introduced during the specimen cutting procedure.

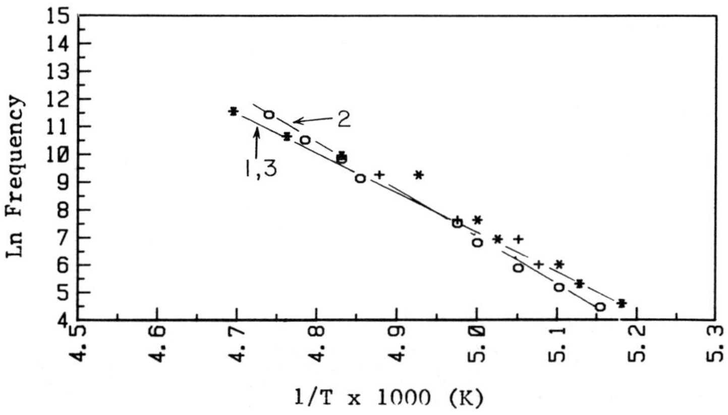

Figure 2. Glass transition map. Key: +, butadiene; ∗, crosslinked butadiene; and ○, crosslinked butadiene exposed to ozone at 375 ppm for 150 min.

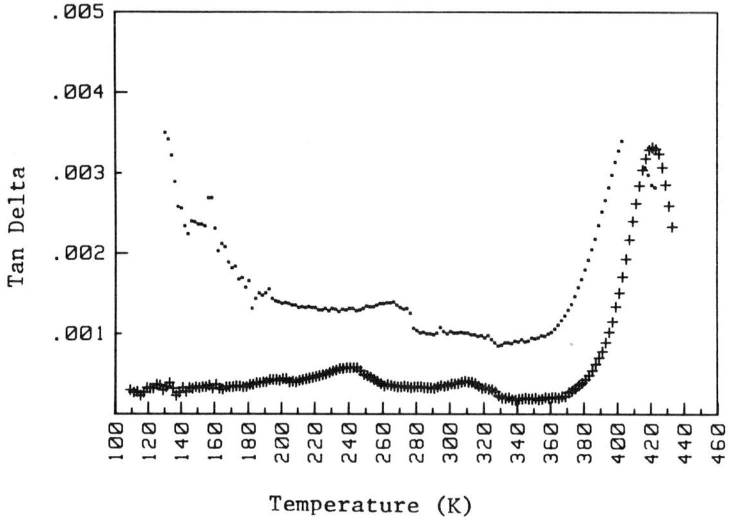

Figure 3. Representative dielectric data for polystyrene taken at 100 kHz. Key: +, untreated; and ●, exposed to ozone at 375 ppm for 150 min.

by the phenyl group in PS with O_2 (11-13) or due to chemical reaction of the styrene repeat unit analogous to addition of a polar group to the phenyl substituent (14-16).

Complex plane analysis (18) of PS dielectric relaxation spectra, (Figure 4) shows that even though the glass transition process has been shifted to a lower temperature, the mechanism involved in the relaxation and the relaxation distribution is essentially unchanged. If crosslinking of the PS has occured it does not appear to be very predominant as compared to chain scission after 75 minutes of ozone exposure. Figure 5 shows the effects of ozone on PS. Degradation of the styrene main chain appears to be occurring as evidenced by the shift in the glass transition to lower temperatures. However, XPS studies indicate that oxygen concentration is not significantly affected at the surface of PS (20).

Tensile

Elongation to failure measured in tensile stress-strain measurements at room temperature (i.e., approximately 25°C) vary significantly as the composition of the polymer is varied from the glassy PS to the elastomeric PB (Figure 6). Also it is clear that the tensile properties of the random copolymer is significantly different from the same composition block copolymer. This difference is due, in part, to the differences in molecular weight (Table 1).

Upon examination of yield stress vs % styrene, it can be easily seen (Figure 7) that increasing the % styrene content also increases the tensile strength of the copolymer. Polystyrene appears to exhibit an interesting phenomenon, when exposed to ozone for increasing lengths of time, the yield stress first increases then decreases with increasing exposure (Figure 8). Sample 310 also appears to exhibit this same phenomenon. Samples with lower % styrene content do not exhibit this affect. This affect may be due to an initial PS crosslinking (10) followed by chain degradation as ozone diffusion continues. XPS studies of these block systems (20) imply that there may be preferential migration of the PS blocks to the surface under these specimen proparation conditions.

Summary

Polystyrene and polybutadiene homopolymers as well as random and block copolymers of these mers have been studied via dielectric relaxation spectroscopy and tensile stress-strain measurements. The results suggest that some block copolymer systems studied have styrene rich surfaces which appear to partially crosslink upon initial exposure to ozone even though surface oxygen concentrations are not significantly affected. After continued exposure these samples appear to then undergo chain scission. Complex plane analysis implies that after degradation

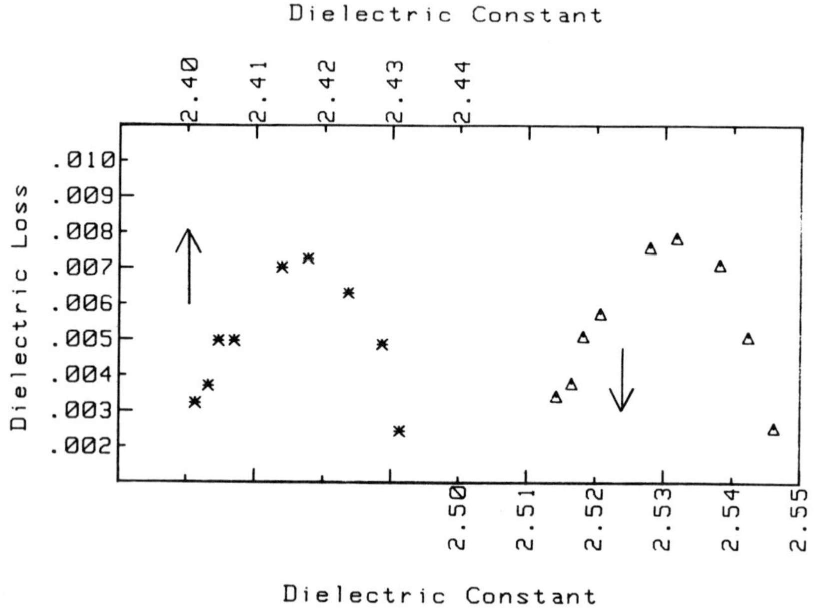

Figure 4. Cole–Cole plot of polystyrene at temperature of 1-kHz glass transition, from dielectric data. Key: ∗, untreated at 397 K; and △, exposed to ozone at 375 ppm for 75 min at 391 K.

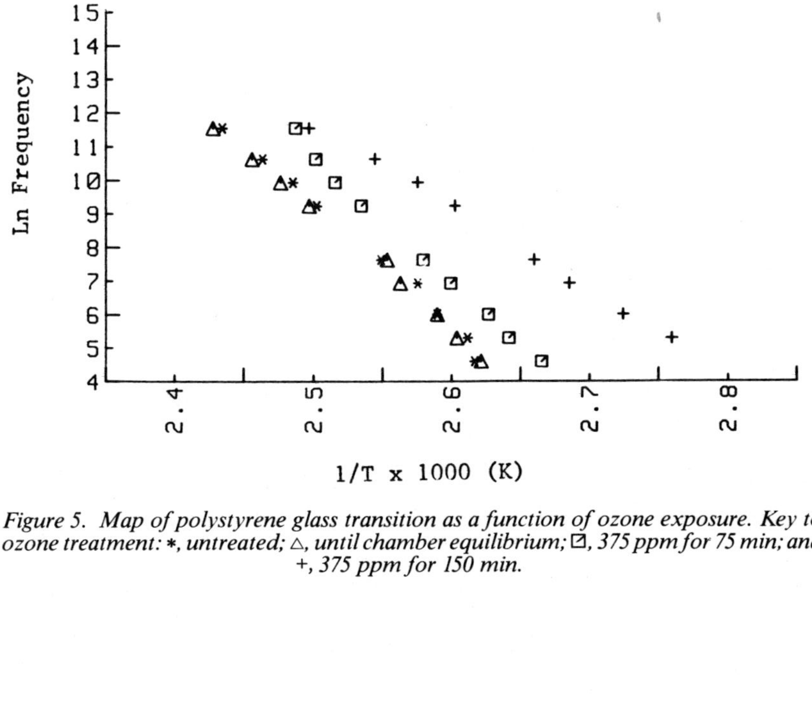

*Figure 5. Map of polystyrene glass transition as a function of ozone exposure. Key to ozone treatment: *, untreated; △, until chamber equilibrium; ◻, 375 ppm for 75 min; and +, 375 ppm for 150 min.*

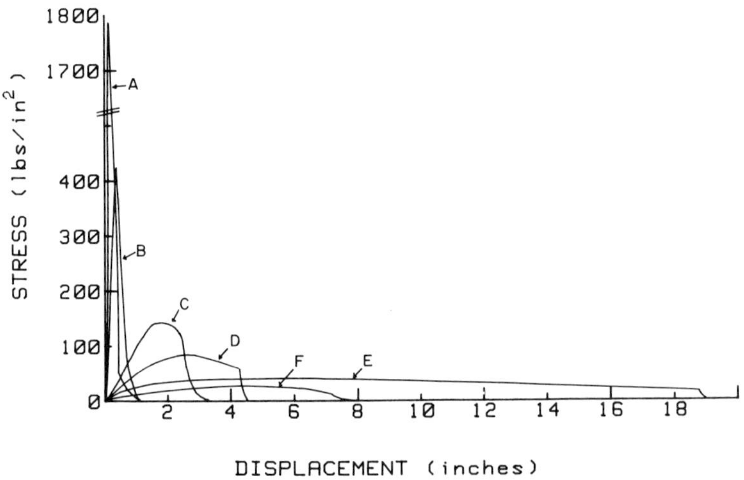

Figure 6. Variations of tensile strength as a function of composition of butadiene (BU) and styrene (ST) polymers and copolymers. Key: A, polystyrene homopolymer; B, 52/48 BU/ST block copolymer; C, 70/30 BU/ST block copolymer; D, 75/25 BU/ST block copolymer; E, 75/25 BU/ST random copolymer; F, butadiene homopolymer.

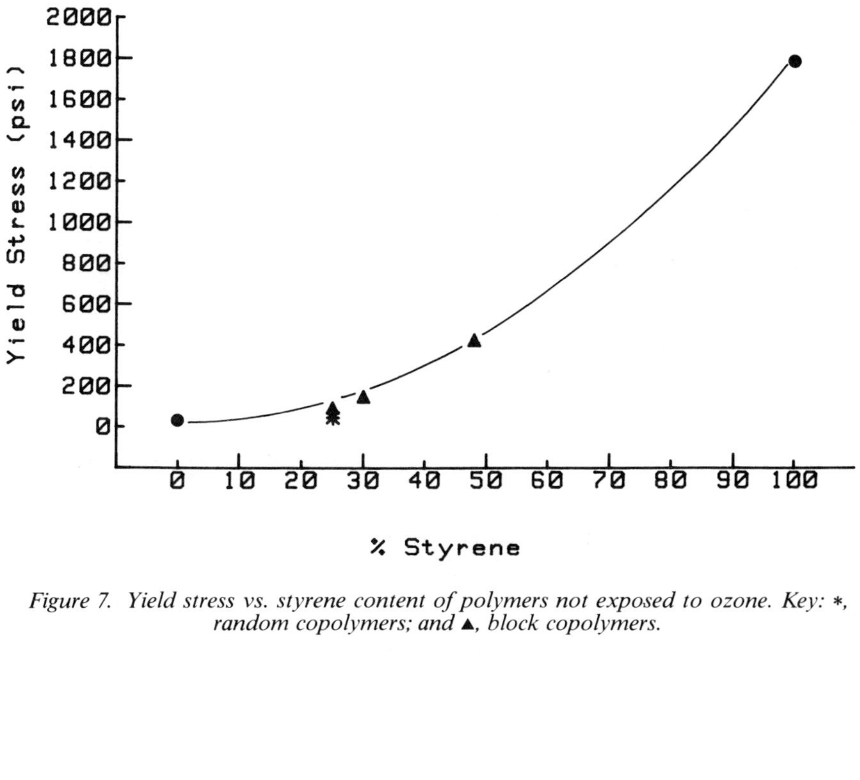

Figure 7. Yield stress vs. styrene content of polymers not exposed to ozone. Key: ∗, *random copolymers; and* ▲, *block copolymers.*

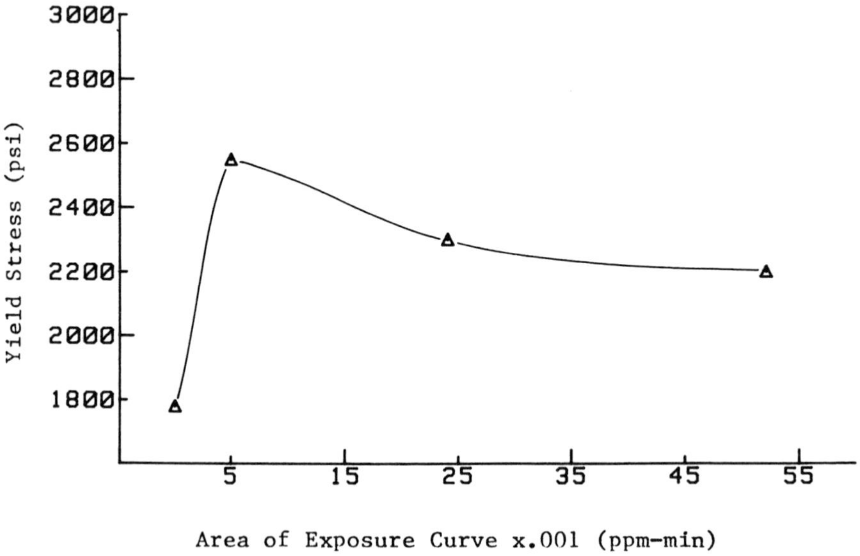

Figure 8. Effects of ozone exposure on yield stress of polystyrene. Area curve units are ppm-min.

has occured, the Tg process for PS is reduced in temperature and occurs by the same mechanism as in non-degraded PS. The butadiene rich materials appear to undergo predominantly chain scission with increases in oxygen concentrations at the surface of the sample. At higher frequencies, butadiene glass transitions appear to shift to lower temperatures with higher activation energies for the Tg process. These changes in bulk physical properties appear to be self-consistent with the surface studies also included in this publication. Current studies are underway using infrared spectroscopy to ascertain products of the ozone reactions as well as lend insight into the diffusion rate. Current work is also underway to understand the effect of deformation during ozone exposure on the rate of degradation.

Attempts at formulating a reciprocity dependence revealed that reciprocity did not hold for these systems. However, it does appear that reaction of ozone with the films is dependent in some way with both ozone concentration and length of exposure.

Acknowledgment

The authors wish to thank the Microstructure of Materials Center of Excellence, College of Engineering, University of Florida for financial support of this research.

Literature Cited

1. Fox, R.B., Price, T.R., and Cain, D.S. in "Interaction of Liquids at Solid Substrates," Alexander, A.L., Ed.; ACS SYMPOSIUM SERIES No. 87, New York, NY, 1968; p.72.
2. Rabek, J.F.; Ranby, B. J. Polymer Sci., Chem. Ed., 1974, 12, 273.
3. Rabek, J.F.; Ranby, B. J. Polymer Sci., Chem. Ed., 1974, 12, 295.
4. Greenwood, R., Weir, N.A. "Ultraviolet Light Induced Reactions in Polymers," Labana, S.S., Ed.; ACS SYMPOSIUM SERIES No.25, New York, NY, 1976; p.220.
5. Weir, N.A. J. Polymer Sci., Chem. Ed., 1978, 16, 13.
6. Ghaffar, A., Scott, A., Scott, G., European Polymer J., 1975, 11, 271.
7. Andries, J.C., Diem, H.E.J. Polymer Sci., Letters, 1974, 12, 281.
8. Thomas, H.R., Beatty, C.L., Limburg, W.M., in press.
9. Saltzman, B.E., Gilbert, N., Analytical Chemistry, 1959, 31, 1914-20.
10. Saito, T. etal., Bull. Chem. Soc. of Japan, 1978, 51, 1153.
11. Nozaki, M., Shimada, K., Okamoto, S., Japan J. Appl. Phys., 1970, 9, 843.
12. Pochan, J.M., Hinman, D.R., J. Polymer Sci., Phys. Ed., 1976, 14, 1871.

13. Froix, M.F., Williams, D.J., Goedde, A.O., Macromolecules, 1976, 9, 354.
14. Irvine,J.D., Work, R.N., J. Polymer Sci., Phys. Ed., 1973, 11, 175.
15. Shimizn, K., Yano, O., Wada, Y., J. Polymer Sci., Phys. Ed., 1975, 13, 2357.
16. Armeniades, C.D., Baer, E., Rieke, J.K., J. Appl. Polymer Sci., 1970, 14, 2635.
17. Negami, S., in "Charactertization of Coatings: Physical Techniques,"Marcel Dekker, Inc. New York, NY, 1976; 2, part II.
18. Cole, K.S., Cole, R.H., J. Phys. Chem., 1941, 9, 341.
19. Hedvig, P., in "Dielectric Spectroscopy of Polymers," Wiley and Sons, Inc., New York, NY, 1977; chapt. 4.
20. Stephens, K., Ammons, M., Batich, C.D., Beatty, C.L., Swartz, W. "XPS Study of the Efffects of Ozone on Various Styrene/ Butadiene Copolymers," contained in this publication.

RECEIVED April 11, 1983

SURFACE EFFECTS OF ENVIRONMENTS

16

x-Ray Photoemission Spectroscopic Study of the Effect of Ozone on Various Styrene/Butadiene Copolymers

K. E. STEPHENS, M. AMMONS[1], C. D. BATICH, C. L. BEATTY and W. SWARTZ[1]

University of Florida, Department of Materials Science and Engineering, Gainesville, FL 32611

A series of butadiene/styrene homopolymers and copolymers were exposed to air-enriched ozone. For butadiene and random copolymers, surface energies quickly increase and oxygen/carbon ratios are large upon initial exposure as evidenced by contact angle measurements and XPS spectra. This is presumed to be simple oxidative degradation of carbon-carbon double bonds. For polystyrene and high % styrene block copolymers, surface energies are increased slightly by ozone but surface oxygen concentrations are not significantly affected. For low percent styrene block copolymers, surface energies approach their highest levels and surface oxygen content their lowest levels upon ozone exposure. This study indicates preferential segregation of styrene at the surface of the block copolymers.

A substantial amount of work has been done on the ozone degradation of polymers. For instance, mechanical properties and adhesion have been shown to degrade sharply during ozone exposure (1-4). Polystyrene (PS) and polybutadiene (PBD) show very different sensitivities to ozone exposure. Since reaction with a gas would occur first at the surface, we planned to study the reactivity of the PBD as affected by copolymerization with styrene. In particular it was of interest to know if random and block copolymers of styrene would show different surface sensitivities because of changes in surface distribution of the groups. For the same group concentration, block structures should segregate PS better. We have carried out contact angle and XPS studies of the various polymers indicated in Table I and have found significant differences in their behavior. Apparently PS serves to protect the surface to a large extent when present in block form. However, styrene/ butadiene copolymers which are randomly polymerized are attached by ozone approximately as efficiently as pure butadiene.

[1]University of South Florida, Department of Chemistry, Tampa, FL 33620.

0097-6156/83/0229-0279$06.00/0
© 1983 American Chemical Society

Table I. Solutions of polymer used to prepare cast films (wt. %)

Sample Code	Sample	Initial Solution	Reprecipitated Solution	Contact Angle Solution
PS	Polystyrene	20	20	5
310	Block 52/48 But/sty	9	9	2.25
308	Block 70/30 But/sty	9	9	2.25
1205	Block 75/25 But/sty	9	9	2.25
1206	Random 75/25 But/sty	9	9	2.25
223	Homopolymer Butadiene	6	6	1.5
XL	Crosslinked Butadiene	6	6	1.5

Experimental

Sample preparation for measurement was done by reprecipitation of the commercial (Solprene) polymers from chloroform solution by adding dropwise to cold methanol. Casting solutions were prepared using HPLC grade chloroform.

Crosslinked samples were made using benzoyl peroxide equivalent to 1% of polymer weight in solution. The solutions were then placed on a roll mill at a slow rate (30 rpm) for 1 hour to facilitate homogeneous dispersion of benzoyl peroxide prior to casting. Sample solutions were poured onto a substrate and protected from dust during chloroform vaporization. After drying overnight in air, samples were placed in vacuum for a minimum of 12 hours to remove residual chloroform. Ozone treated samples were exposed to an air flow of 1.5 liters/min. containing 375 ozone for 25, 75 or 150 minutes in a bell-jar. Ozone generation was achieved using a P.C.I. Model #C2P-3C high-voltage ozone generator and ozone output was calibrated by oxidation of standard KI (pH 6.74) solutions with sodium thiosulfate titration.

Upon opening the ozone chamber to introduce samples, a concentration gradient was produced. For this reason and the subsequent time involved for equilibrium to be re-established, exposure time was not used as the only guage of exposure in reciprocity measurements. Under these circumstances, a factor including the area under the concentration vs time curve appears to be a better measure of exposure. This area factor was kept constant over a range of concentration conditions for the reciprocity measurements. All samples were placed in vacuum for at least 4 hours after ozone exposure to remove residual gas adsorbed to the surface. The substrates for contact angle samples were glass slides with final solutions diluted 3:1 with $CHCl_3$ to facilitate a more even surface upon drying.

Residual solvent was removed by placing slides in a vacuum for 2 days prior to measurement or ozone exposure. Measurements were made using a Rame' Hart, Inc. Model 100-00-00 NLR Goniometer. Five liquids were measured for each sample and data for at least four of these were included for calculations of γ_c. Six measurements of angle were made for each liquid on each sample. The substrate used for the XPS samples was aluminum foil. XPS spectra were obtained on a McPherson 36 instrument and were measured at normal incidence.

Contact Angle Measurement and Critical Surface Energy

Example of typical data and data scatter derived from contact angle measurements are shown in Figure 1. It can be seen (Figure 2) that the effects on surface energy of ozone exposure are

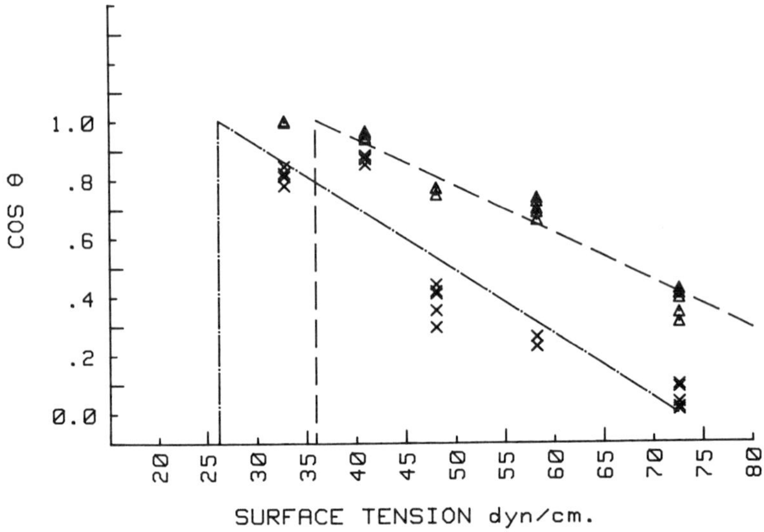

Figure 1. Representative contact angle and critical surface energy data. Key: ▲, untreated 52/48 butadiene/styrene block copolymer; and ×, butadiene exposed to ozone at 375 ppm for 150 min and crosslinked with 1% benzoyl peroxide.

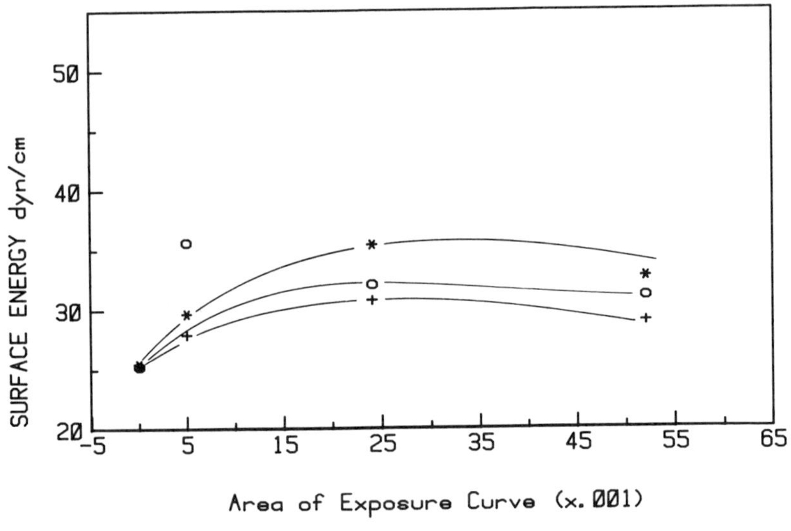

*Figure 2. Effects of ozone exposure on critical surface energy. Exposure curve units are ppm-min. Key: *, butadiene; ○, crosslinked butadiene; and +, 75/25 butadiene/styrene random copolymer.*

decreased when 25% styrene is added randomly to the chain structure. This is presumably due to the greater resistance of styrene to ozone vs carbon-carbon double bonds in the butadiene segments. Therefore, even if some styrene is degraded, or oxidized, or crosslinked, the reaction rate is assumed to be slower for styrene due to it's aromatic nature. In cases where the block copolymer is present, it shows a larger surface free energy than the random material. From the surface energies of the pure components this suggests that more PS is at the surface in the block copolymer than in the random copolymer. This data should relate to the chemical composition of the top 5 Å of material and suggests that all treated surfaces are similar. In examining the block copolymers, all samples tested showed an initial, significant increase in surface energy followed by a region of slow change. As the percent styrene was increased this region of slow change appeared to move from slight positive to slight negative slope indicating more resistance to surface energy increases with increasing percent styrene.

Since the initial increase in surface energy becomes much less significant as percent styrene is increased, it appears that if styrene is reacting, it is producing products with lower surface energy than a reaction with butadiene segments, or the styrene segments are resistant and acting as a shield to prevent reaction of butadiene segments.

Upon examination of Figure 3 the butadiene homopolymer surface energy does not appear to be as strongly affected as the 25% styrene block copolymer. This realization alone would tend to discount the theory of styrene simply shielding ozone from the butadiene segments. The effect of ozone treatment on the change in surface energy appears to relate to the styrene content of the polymers studied.

In looking at surface energies vs styrene content, (Figure 4) it can be seen from this data that as ozone exposure is increased, the deviation of (25-30%) styrene block copolymers from the other samples becomes more pronounced. This implies that butadiene and styrene are both involved in some reaction with ozone (or each other) and this reaction is enhanced at these concentrations of block copolymer. Figures 2 and 4 would tend to imply that this phenomenon does not occur with the random copolymer or homopolymers nor with block copolymers with more than 30% styrene content.

XPS Study

Typical XPS spectra are shown in Figure 5. This technique samples a greater depth (about 30 Å) and is able to give some detailed chemical state information about the composition (3). Survey scans of the polymer reveal trace amounts of silicon contaminant, but represent only about 2% of the total surface area and can be subtracted from the final spectra. Silicones are

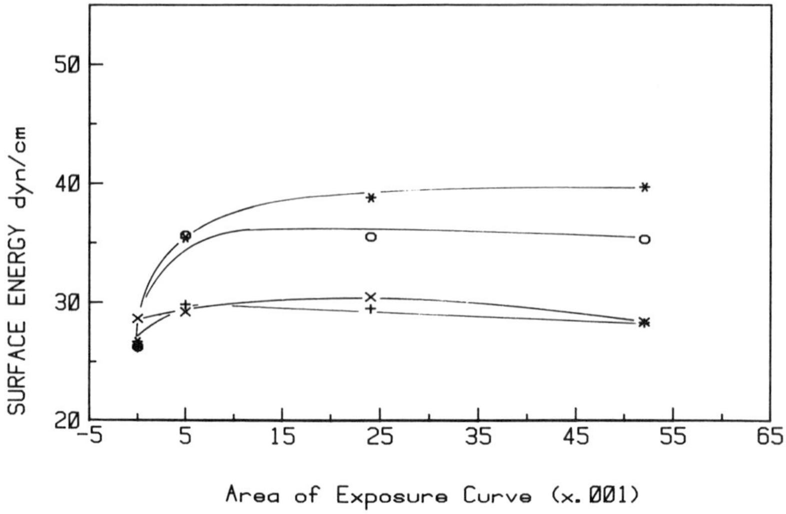

*Figure 3. Effects of ozone exposure on critical surface energy of butadiene (BU) and styrene (ST) polymers and copolymers. Exposure curve units are ppm-min. Key: *, 75/25 BU/ST block copolymer; ○, 70/30 BU/ST block copolymer; +, 52/48 BU/ST block copolymer; and ×, polystyrene homopolymer.*

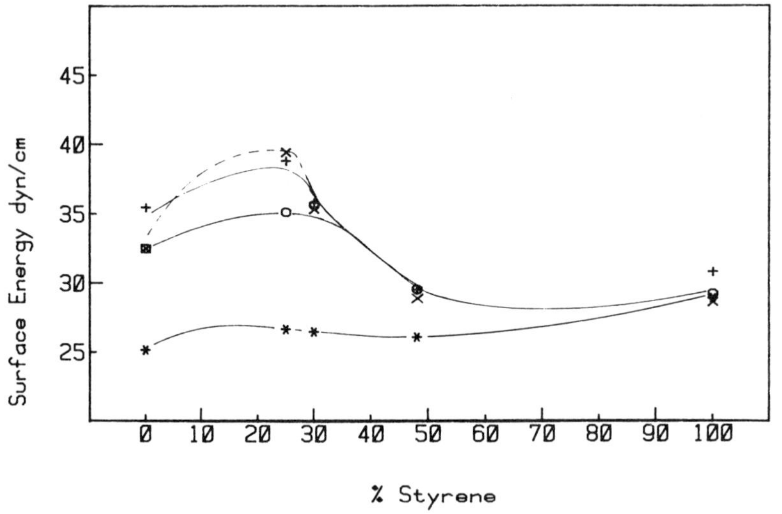

*Figure 4. Effects of ozone exposure on critical surface energy of block copolymers and homopolymers as a function of bulk percent styrene content. Key to ozone treatment: *, untreated; ○, until chamber equilibrium; +, 375 ppm for 75 min; and ×, 375 ppm for 150 min.*

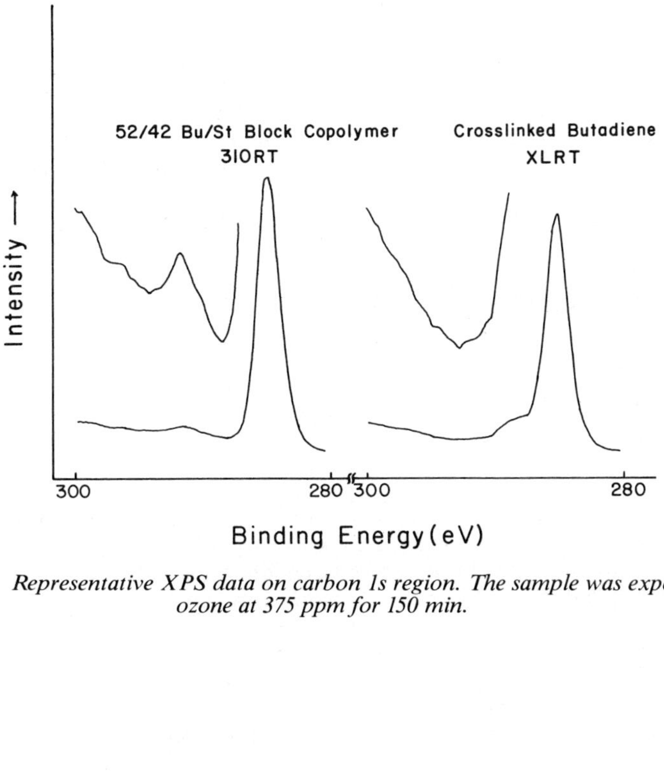

Figure 5. Representative XPS data on carbon 1s region. The sample was exposed to ozone at 375 ppm for 150 min.

common contaminants in surface films. Some of the polymers initially had oxygen present on the surface to the extent of about 10%. This occurs from air oxidation during processing and has been found for most commercial hydrocarbon polymers prepared in the laboratory or in industrial processes. Upon treatment with ozone, polymer oxygen levels increase sharply for the pure PBD and also for the random copolymers (Figure 6). This suggests that in the random copolymer a significant fraction of butadiene is being reacted with ozone in the top 30 A. It is surprising that the oxygen concentration on the surface actually decreases for several of the polymers (Figure 7). This may be due to non-chain cleavage oxidation (e.g., ketone) being removed by chain degradation and volatization. Gravimetric analysis is currently being considered to answer this question. In this and other publications (5) we have implied that under these experimental conditions, styrene was preferentially migrating to the surface in the block copolymer. Examining the oxygen carbon ratio for the 52% butadiene block copolymer, (#310), appears to show instances where some samples acted as butadiene and others acted as styrene. The reason for this anomaly is not understood at the present and certainly deserves more study, but it is probably related to solvent casting techniques.

More chemical information can be obtained from looking at typical carbon spectra which are shown in Figure 5. Initial (untreated films) spectra show a single peak with the small satellite structure if significant PS is present. Polybutadiene and also the random copolymers show a large shoulder appearing at 288 electron volts after ozone exposure. This corresponds to carbon with one oxygen attached and could be due to an alcohol, hydroperoxide, carbonyl group, ozonide, or ether-type linkage. In contrast, the ozone exposed polymers which contain significant polystyrene present as a block do not show this shoulder on the carbon 1s peak. After exposure, some of the PS films showed the development of a satellite structure at about 292 eV. This peak could correspond to a UV-type transition giving a satellite shake-up structure or it could correspond to a carboxylic acid formed in low amounts. The sharp difference between the random and block 75/25 BD/S copolymers are especially interesting. This difference suggests that a relatively small amount of polystyrene, if present as a block additive can surface segregate and restrict reaction somewhat with the butadiene present. This is consistent with observed ozone/polymer reactions which show PS undergoing cross-linking and chain-scission leading to a broadened molecular weight distribution while PBD shows more scission. These results are related to mechanical property and other measurements on these same films (5).

When surface energy and oxygen-carbon ratios are considered as dependent variables, apparent relationships appear at higher exposure times. Figure 8 suggests that low percent styrene (30% or less) block copolymers have a high surface energy for low O/C ratio. Polystyrene and higher percent styrene block copolymers

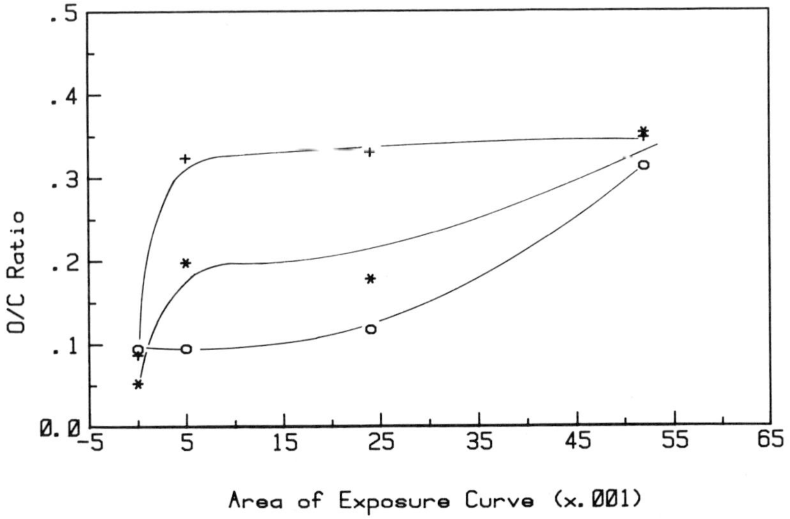

*Figure 6. Effects of ozone exposure on oxygen/carbon ratio. Exposure curve units are ppm-min. Key: *, butadiene; ○, crosslinked butadiene; and +, 75/25 butadiene/styrene random copolymer.*

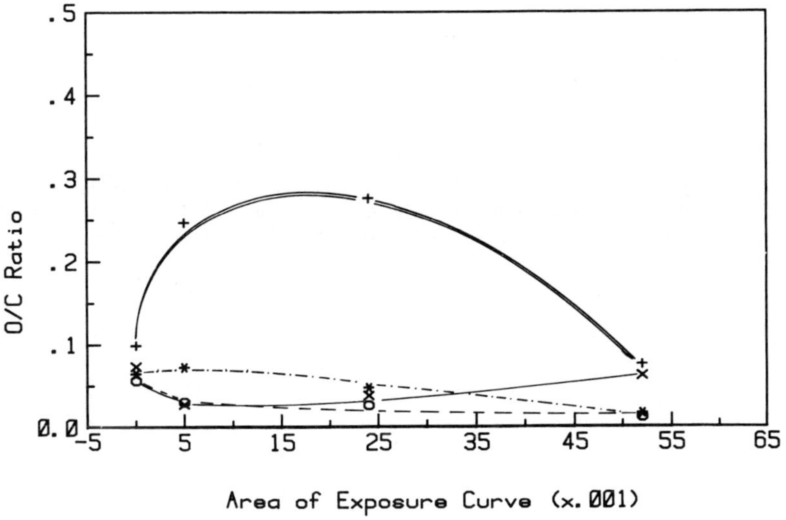

*Figure 7. Effects of ozone exposure on oxygen/carbon ratio of butadiene (BU) and styrene (ST) polymers and copolymers. Key: *, 75/25 BU/ST block copolymer; ○, 70/30 BU/ST block copolymer; +, 52/48 BU/ST block copolymer; and ×, polystyrene homopolymer.*

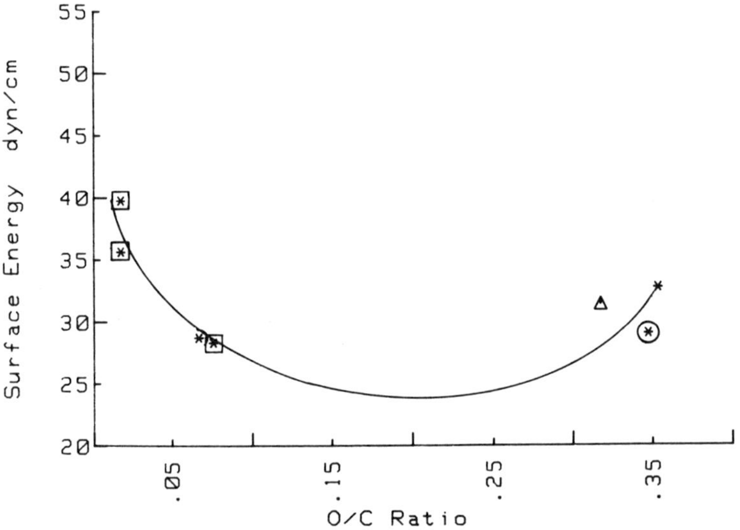

Figure 8. Critical surface energy as a function of oxygen/carbon ratio for butadiene/styrene copolymers and homopolymers after ozone exposure at 375 ppm for 150 min. Key: △, crosslinked butadiene; ○, random copolymers; and ⊡, block copolymers.

are very similar on the surface, but behave differently than lower percent styrene block copolymers. For these low percent styrene block copolymers, the ratio is increased significantly, while the surface energy is reduced. Finally, butadiene and random copolymers are affected differently than either group above and consistantly produce high values of the O/C ratio.

Results from reciprocity measurements for XPS and contact angle have at this point been inconclusive. However, physical properties data (5) appears to confirm that there may be an optimum concentration of ozone above which reaction rates may not necessarily increase, and there may be a limiting exposure time below which diffusion into the bulk does not occur.

Summary

A series of butadiene/styrene homopolymers and copolymers were exposed to air enriched ozone for various exposure times. Evidence suggest that at least three reaction mechanisms may occur with ozone and the polymers. For butadiene and random copolymers, surface energies quickly increase and oxygen/carbon ratios are high upon exposure. This is presumed to be simple oxidative degradation of butadiene carbon-carbon double bonds.

For polystyrene and high percent styrene block copolymers, surface energies are increased slightly but surface oxygen concentration is not significantly affected. This is postulated to be initial styrene crosslinking and some non-cleavage oxidation, followed by chain degradation and some volatilization, of oxidative products.

For low percent styrene block copolymers surface energies approach their highest levels and surface oxygen content the lowest levels upon exposure to ozone. This mechanism appears to be different than the previous mechanisms and occurs only when styrene blocks are present in low concentrations compared to butadiene and also appear to reside preferentially at the surface. Infrared spectroscopy and gravimetric studies are currently being considered as further work in an attempt to answer some questions posed by this work.

Acknowledgment

The authors would like to thank the Microstructure of Materials Center of Excellence for financial support of this research.

Literature Cited

1. J. Rabek, "Comprehensive Chemical Kinetics", ed. by C. Bamford and C. Tipper, Vol. 14, Degradation of Polymers, Elsevier, 521 (1975).

2. B. Cherry, "Polymer Surfaces," Cambridge University Press, 28 (1981).
3. "Polymer Surfaces" ed. by D. T. Clark and W. Feast, John Wiley and Sons Chpt. 16 and 18 (1978).
4. T. Saito et. al., Bull. Chem. Soc. of Japan, 51, 1153 (1978).
5. K. Stephens and C. L. Beatty, "Effect of Ozone Exposure on the Physical Properties of Butadiene and Styrene/Butadiene Copolymers", contained in this publication.

RECEIVED April 11, 1983

17

Environmental Corrosion of Polymers

Causes and Main Reactions

B. RÅNBY and J. F. RABEK

The Royal Institute of Technology, Department of Polymer Technology, Stockholm, Sweden

Air pollutants such as particulate matter, sulfur oxides, nitrogen oxides, hydrocarbons and photochemical oxidants such as ozone, atomic oxygen, singlet oxygen and peroxyacetylnitrate may cause serious corrosion of polymeric materials: rubber, plastics and fibres. The paper reviews the problem of environmental corrosion of polymers considering the effects and mechanisms of corrosion by the particular air pollutants. Protection and stabilization of rubber and plastics against these degradation processes is of great importance from an industrial point of view. It also affects the saving of raw materials and energy. Polymers more resistant to environmental corrosion mean longer lifetime which decreases the necessary production.

Air pollution is usually defined as the presence in the outdoor atmosphere, of substances put there directly or indirectly by an act of man, in amounts which are detrimental to health and safety or interfere with the full use of materials or property, e.g. made from polymeric materials.[1-4] Man is not the only agent able to pollute the atmosphere. There are many natural processes that do so such as pollination of plants, volcanic emptions, dust stormes and forest fires. There are also secondary pollutants in the atmosphere, formed in the air from primary pollutants, due to acts by man, e.g. smoke, industrial pollutants and photochemical smog formed in the air from substances emitted from automobile exhaust and other sources.

The major air pollutants are

Particulate matter. There is a large variety of such materials in the air, primarily solid matter rather than liquid. The particles come from a variety of sources, they are of different size, shape, color, texture, and chemical composition. They can remain suspen-

0097-6156/83/0229-0291$06.00/0
© 1983 American Chemical Society

ded in the air for periods ranging from a few seconds to a few years.

Sulfur oxides, in the form of sulfur dioxide (SO_2) sulfur trioxide (SO_3), sulfuric acid (H_2SO_4) and sulfates (SO_4). Most of these pollutants are emitted to the atmosphere as sulfur dioxide, which is chemically converted by oxidation to SO_3 and sulfates in the air. Both particulate matter from smoke and sulfur oxides are known to be components in the smogs of London. They are mainly the result of combustion of coal.

Carbon monoxide (CO) is produced primarily by the combustion of gasoline in automobile engines.

Nitrogen oxides, in the form of nitric oxide (NO) and nitrogen dioxide (NO_2). Both gases are by-products from combustion of fuel, at high temperatures causing chemical combinations of nitrogen and oxygen. The products are ingredients in the subsequent production of photochemical smog.

Hydrocarbons, which are a large class of chemical compounds mainly formed by incomplete burning of fuel and involved in the smog production.

Photochemical oxidants, which are chemically or photochemically formed in the air. The stable oxidants are ozone (O_3), and peroxyacetyl nitrate (PAN), the unstable are atomic oxygen (O) and singlet oxygen ($^1O_2$).

Photochemical oxidants are mainly produced during power generation, in industrial operations and are also from natural sources. The singlet oxygen (3-6) is to a large extent related to automobile exhaust, which is the main cause of the photochemical smog in the Los Angeles area.

This classification (1-6) of air pollutants has been legally established by the Environmental Protection Agency (USA) as the National Ambient Air Quality Standards under the terms of the Clean Air Act.

Air pollutants (1-6) are mainly the result of burning of various types of fuels: coal, oil, and gas for heating of building, generation of electricity and gasoline for transport etc. The use of gasoline and diesel fuel produces some of the particulate matter, much of the hydrocarbons and nitrogen oxides, and hence much of the photochemical smog, and almost all the carbon monoxide. Coal and oil combustion in industry and local heating produces most of sulfur oxides and particulates. Natural gas is the "cleanest" fuel. Combustion of this gas produces some nitrogen oxides, but not as much as the burning of gasoline in automobile engines.

Industrial processes produce only half as much particulate matter as fuel combustion, one fourth of the SO_2, and smaller portions of the other air pollutants. The pollution caused by industry is less significant than pollution from fuel combustion in automobiles.

Having defined the various air pollutants, we now consider briefly the types of effects which are possible in environmental corrosion of polymers.

Mechanical corrosion of polymers by particulate matter

The particles in the air over an urban area are of a variety of sizes, shapes and chemical composition, ranging from tiny, spherical metal particles from metallurgical fumes to huge, porous conglomerates of sooty carbon, soil particles, fly ash, and fly dust of all types.[5] The size and shape of the particles will almost totally determine the surface corrosion of polymers, plastics, the polymer in paints and lacquers. The important question is how these particles behave in the air, how far and how fast the wind carry them, and what effects they can have on polymeric materials.

Particles larger than about ten microns are emitted into the air primarily by physical processes, such as grinding and abrasion, and flash from fuel combustion. The particles tend to settle out onto the ground (or on the surface of polymeric materials) rather rapidly. These are the particles gathered as dustfall. In environmental pollution control the dustfall material is weighed and reported as tons per square mile per month or, in metric units, as grams per square meter per day. Dustfall plays a fundamental role in the abrasive corrosion of polymer surfaces but it also decreases the photochemical processes (degradation, crosslinking and oxidation processes) by scattering and absorption of harmful sun irradiation.

Unlike dustfall, particles smaller than about ten microns in size remain suspended in the air for long periods of time. Turbulent motion in the air is able to keep them aloft more or less permanently.

These particles may also cause a slow and very fine surface corrosion of the polymeric materials. They also absorb and scatter sun irradiation, but they do not protect polymer surfaces against harmful exposure to sun irradiation like large particles.

The finest particles, those below 0.1 micron in size, are so small that they have no visible effect on the corrosion of a polymer surface. The cleanest air almost always contains at least several hundred such particles per cubic centimeter, and in polluted urban air this may reach 100,000 per cubic centimeter. Even at these large numbers, the small particles don't have corrosion effect and they can be ignored.

There is a special kind of particulate matter so called "black smoke" that cover polymer surfaces very tightly and is difficult to remove. Black smoke may contain stable unpaired electrons (free radicals) which may play a specific role in photostabilization of polymeric materials by scavenging free radicals formed from photolysis of polymers.

The amount of particulate matter emitted into the atmosphere amounts to millions of tons each year as reported for USA (Table I). Corresponding data for Sweden are not available.

Table I. Annual emissions of particulate matter (estimates for 1970 in USA)[6]

Source category	Emissions, millions of tons per year	Percent of total
Fuel combustion		
Transportation		
Gasoline motor vehicles	0.3	1.1
Diesel aircraft, trains, vessels	0.3	1.2
Off-highway vehicles	0.1	0.4
Total transportation	0.7	2.7
Stationary sources		
Coal	5.6	21.5
Fuel Oil	0.4	1.5
Natural gas	0.2	0.8
Wood	0.6	2.3
Total stationary source	6.8	26.1
Total fuel combustion	7.5	28.8
Industrial processes	13.3	51.0
Agricultural burning	2.4	9.2
Solid waste disposal	1.4	5.3
Miscellaneous	1.5	5.7
Total	26.1	100.0

It is apparent that the bulk of this particulate matter comes from fuel combustion sources, primarily electric-power plants and heating boilers, and from industrial processes. This fact should be taken into consideration of surface corrosion and/or covering by the dustfall levels in the old industrialized cities, whereas values approaching 2000 tons/mi^2 month are found in heavy industrial regions. In the outermost suburbs of a big city the dustfall would be 5 tons/mi^2 month.

At relatively low velocities, small particles hitting a plastic material with impact will abrade the surface and alter its properties. At some critical velocity, e.g. of a vehicle or blowing air the target surface is penetrated and the particles are deformed or break up as they pass into the material. Penetration of the plastic surfaces by the particulate matter is usually accompanied by the formation of a deep cylindrical cavity of the target plastic.

At hypervelocities (e.g. in hypersonic aircraft or rockets), the impact of particle matter on a plastic surface transfers a tremendous amount of kinetic energy within a small region. This energy is partitioned into heat, light, explosive mass ejection, and shock wave generation[8]. The temperatures are of sufficient magnitude to produce localized melting thermal degradation and even complete pyrolysis of the polymer.

The magnitude and modes of damage to plastics by particulate matter impact is dependent upon a number of parameters, such as mass, size, shape, velocity, physical state and trajectory of the particle. The density, hardness, strength, sonic velocity, crystal structure, and other properties of the materials are also significant. Soft and rubbery plastics have higher spallation threshold and less crater damage than rigid plastics. Greater penetration frequency is also obtained with the impact of small particles on soft plastics or ablating plastics containing a molten surface layer. Plastic composites containing many interfaces are better able to withstand impulsive stress of short magnitude, because they distribute energy over a wide area.

Punctures, craters, cracks and other types of mechanical corrosion which are noted in particle impact on plastics may cause serious degradation and even result in catastrophic failure of the material.

Corrosion of polymers by sulfur oxides

Sulfur oxides (SO_x) include not only SO_2 but also sulfur trioxide (SO_3). Sulfuric acid mist and sulfates may also be derived from sulfur oxides, but they are commonly not defined as part of SO_x. Most of the sulfur contamination is emitted in the form of sulfur dioxide with about 1 to 3 percent of sulfur trioxide mixed in.

In the air, SO_2 reacts with oxygen, ammonia, and other compounds, including the water vapor present in air, to form sulfuric acid mist, liquid drops of concentrated acid, as well as various other sulfates, The acid aerosoles attack polymer surfaces and cause lustrelessness of the surfaces. The main reaction of SO_2 with hydrocarbons is formation of sulphonic acids (RSO_2H)[9]. On the other hand SO_x shows photochemical reactivity under sun irradiation (up to 388 nm). The excited form of SO_2 reacts with several polymers such as polyolefins, polystyrene, rubbers, nylons and acrylates[10-19]. The main reactions are photosulfonation accompanied by free radical oxidation and chain scission and/or crosslinking reactions. It has also been reported that the mixture SO_2-O_2 absorbs light efficiently below 300 nm and a substantial amount of singlet oxygen $^1O_2$ is produced with a quantum yield of ϕ <0.02[20].

Less recognized but also important is SO_2 damage to paper and leather. Paper products produced from about 1750 on are embrittled by sulfur compounds converted to sulfuric acid which causes hydrolysis. Thus old books which are not stored in sealed cases undergo a gradual deterioration. Leather bindings are similarly affected.

In addition to SO_2 damage of synthetic and natural polymers, another destructive mechanism called "filiform" corrosion should be considered. This particular type of underfilm electrochemical corrosion (observed in polymer coated metal surfaces) takes place along with polymer damages as threadlike patterns which form under high humidity conditions and is accelerated by the presence of SO_2 and inhibited by carbon dioxide.

The problem is serious because the amount of sulfur oxides emitted into the atmosphere has increased rapidly and amounts to millions of tons each year (Table II).

Table II. Annual emissions of sulfur oxides (estimates for 1970) in USA[6]

Source category	Emissions, millions	Percent
Fuel combustion		
Transportation		
Gasoline motor vehicles	0.2	0.6
Diesel, aircraft, trains, vessels	0.6	1.8
Off-highway vehicles	0.2	0.6
Total transportation	1.0	3.0
Stationary sources		
Coal	22.2	65.4
Fuel Oil	4.2	12.4
Natural gas	-	-
Wood	0.1	0.3
Total stationary source	26.5	78.1
Total fuel combustion	27.5	78.1
Industrial processes	6.0	17.7
Agricultural burning	0.1	0.3
Solid waste disposal	0.1	0.3
Miscellaneous	0.2	0.6
Total	33.9	100.0

In Sweden the total amount of sulfur oxides deposited from the atmosphere was measured for 1970: 33.4×10^3 ton and for 1980: 29.0×10^3 ton. The concentrations varied and amounted anually to 1687 mg/m^2 in Goteborg, 989 mg/m^2 in Stockholm and 73 mg/m^2 in Bohus (data for 1980)[21].

The acid build-up in lakes and rivers has decreased the pH to 4.1 - 6.2 in Scandinavia. This is another example of the effects of pollutants. It is mainly the result of long-distance transport of sulfur compounds. The increasing acidity has serious biological effects. It also causes surface corrosion of plastic boats which are estimated to be 500,000 in Sweden.

Reduction of forest growth is also suspected in Sweden because of the gradual acidification of the forest land.

Corrosion of polymers by nitrogen oxides

The amount of nitrogen oxides emitted into the atmosphere amounts to millions of tons each year (Table III). In Sweden the total amount of nitrogen oxides emitted was measured for 1970: 8.4×10^3 ton and for 1980: 9.9×10^3 tons. The annual concentrations varied from 578 mg/m^2 in Göteborg, 245 mg/m^2 in Stockholm to 26 mg/m^2 in Abisko (data for 1980)[21].

Table III. Annual emissions of nitrogen oxides (estimates for 1970 in USA)[6]

Source category	Emissions, millions of tons per year	Percent of total
Fuel combustion		
Transportation		
Gasoline motor vehicles	7.8	34.2
Diesel, aircraft, trains, vessels	2.0	8.8
Off-highway vehicles	1.9	8.3
Total transportation	11.7	51.3
Stationary sources		
Coal	3.9	17.1
Fuel Oil	1.3	5.7
Natural gas	4.7	20.6
Wood	0.1	0.4
Total stationary source	10.0	43.8
Total fuel combustion	21.7	95.1
Industrial processes	0.2	0.9
Agricultural burning	0.3	1.3
Solid waste disposal	0.4	1.8
Miscellaneous	0.2	0.9
Total	22.8	100.0

The main reaction of NO_2 with polymers (such as polyolefins, polystyrene, nylon, rubbers) is abstraction of a hydrogen from the polymer molecules (PH)[13,16,22-36].

$$PH + \cdot NO_2 \xrightarrow{h\nu} P^{\cdot} + HNO_2 \qquad (1)$$

The polymer alkyl radical ($P^{\cdot}$) formed can further react with molecular oxygen and produce polymer peroxy radicals (cf. Section 7).

Nitroxy radicals may also react with unsaturated bonds in polymers according to the reactions[37]:

$$\cdot NO_2 \longrightarrow \dot{N}O + O \quad (2)$$

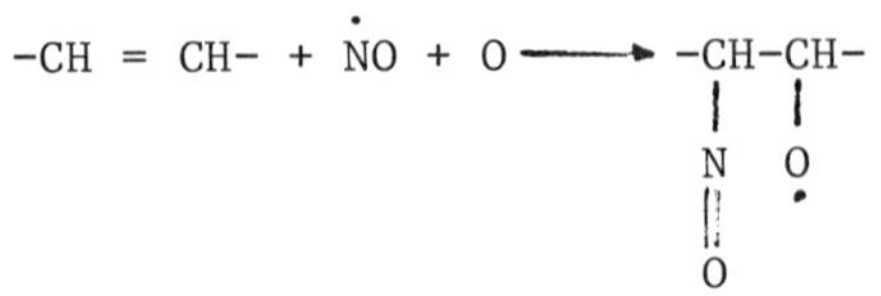

(3)

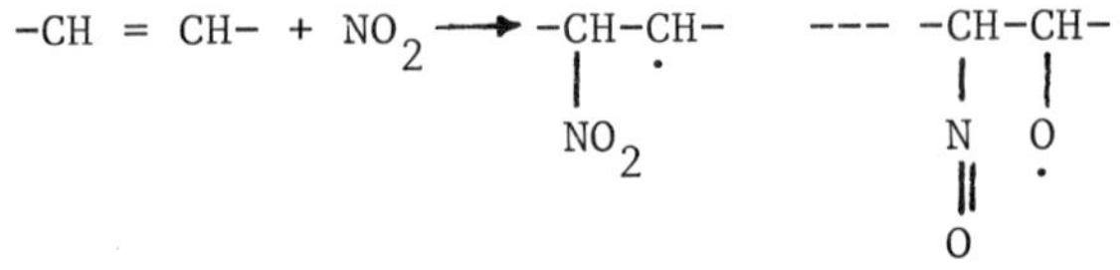

(4)

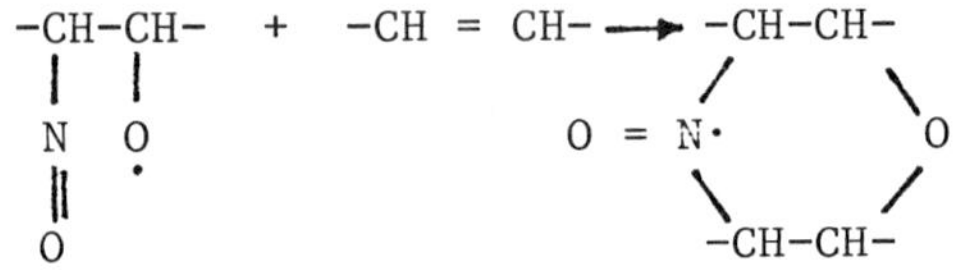

(5)

Formation of nitroxide and iminoxy radicals is enhanced by ultraviolet radiation. This indicates that the photochemical dissociation of NO_2 into nitric oxide and oxygen atoms plays an important role in the reaction with double bonds.

Photochemical smog

The photochemical smog in the atmosphere is a result of chemical and photochemical reactions in the air between oxides of nitrogen, oxygen and hydrocarbons[38-40]. The photochemistry of smog involves very complicated reactions which are intensively investigated by several laboratories round the world.

In the absence of hydrocarbons, the cycle initiated by NO_2 consists of three reactions:

1. Nitrogen oxide has an unpaired electron and is a brown gas. It absorbs solar energy and is dissociated into one molecule of NO and one free oxygen atom,

$$\cdot NO_2 + h\nu \longrightarrow NO\cdot + O \quad (6)$$

2. The free oxygen atom reacts with an oxygen molecule (O_2) to form ozone (O_3),

$$O + O_2 \longrightarrow O_3 \quad (7)$$

Ozone absorbs solar energy and is easily photolysed into atomic oxygen (O), singlet oxygen ($^1O_2$) and molecular oxygen (O_2):

$$O_3 + h\nu\ (11{,}800\ \text{Å}) \longrightarrow {}^3O_2(^3\Sigma_g^-) + O \qquad (8)$$

$$O_3 + h\nu\ (6200\ \text{Å}) \longrightarrow {}^1O_2(^1\Delta_g) + O \qquad (9)$$

$$O_3 + h\nu(4600\ \text{Å}) \longrightarrow {}^1O_2(^1\Sigma_g^-) + O \qquad (10)$$

$$O_3 + h\nu\ (4000\ \text{Å}) \longrightarrow {}^3O_2(^3\Sigma_g^-) + O \qquad (11)$$

$$O_3 + h\nu(2600\ \text{Å}) \longrightarrow {}^1O_2(^1\Sigma_g^+) + O \qquad (12)$$

${}^3\Sigma_g^-$ is molecular oxygen in the ground state (triplet).

${}^1\Delta_g$ and ${}^1\Sigma_g^+$ are two forms of excited states (singlets).

3. An ozone molecule may also react with a NO molecule produced in reaction (6), which forms nitrogen dioxide again:

$$O_3 + NO \longrightarrow O_2 + NO_2 \qquad (13)$$

In the presence of hydrocarbons these reactions become more complex:

NO_x cycle:

$$NO_2^{\cdot} + h\nu \longrightarrow NO\cdot + O \qquad (14)$$

$$O + O_2 + M \longrightarrow O_3 + M \qquad (15)$$

$$O_3 + NO\cdot \longrightarrow NO_2^{\cdot} + O_2 \qquad (16)$$

$$O_3 + NO_2^{\cdot} \longrightarrow NO_3^{\cdot} + O_2 \qquad (17)$$

$$NO_3^{\cdot} + NO_2^{\cdot} \xrightarrow{H_2O} 2HNO_3 \qquad (18)$$

(17)–(18): produces nitric acid

$$NO\cdot + NO_2^{\cdot} \xrightarrow{H_2O} 2HNO_2 \quad \text{nitrous acid} \qquad (19)$$

$$HNO_2 + h\nu \longrightarrow NO^{\cdot} + \cdot OH \qquad (20)$$

$$CO + \cdot OH \longrightarrow CO_2 + HO_2^{\cdot} \quad \text{CO effect produces an OH chain} \qquad (21)$$

$$HO_2^{\cdot} + NO^{\cdot} \longrightarrow NO_2^{\cdot} + \cdot OH \qquad (22)$$

$$HC\cdot + O_2 \longrightarrow RO_2\cdot \qquad (23)$$

$$HC\cdot + O_3 \longrightarrow RO_2\cdot + RCHO \quad \text{aldehyde and peroxy radicals} \qquad (24)$$

$$HC^{\cdot} + OH + O_2 \longrightarrow RO_2^{\cdot} + RCHO \qquad (25)$$

$$HC\cdot + RO_2 \longrightarrow RO_2\cdot + RCHO \qquad (26)$$

$$HO_2\cdot + NO \longrightarrow NO_2 + \cdot OH \quad \text{oxidation of NO to } NO_2 \qquad (27)$$

$$RO_2 + NO_2 \longrightarrow PAN \quad \text{production of peroxy-acetylnitrate} \left(\overset{O}{\overset{\|}{RCOONO_2}} \right) \qquad (28)$$

Several oxygen species such as atomic oxygen (O), ozone (O_3) and singlet oxygen ($^1O_2$) which are formed in a photochemical smog are very reactive with polymers (cf. Section on oxidation).

Photosensitized reaction by hydrocarbons

The amounts of hydrocarbons emitted into the air are millions of tons each year as estimated for USA. (Table IV). Corresponding data for Sweden are not available.

Table IV. Annual emissions of hydrocarbons (estimates for 1970 in USA)[6]

Source category	Emissions, millions of tons per year	Percent of total
Fuel combustion		
Transportation		
Gasoline motor vehicles	16.6	47.6
Diesel, aircraft, trains, vessels	0.9	2.6
Off-highway vehicles	2.0	5.7
Total transportation	19.5	55.9
Stationary sources		
Coal	0.2	0.6
Fuel Oil	0.1	0.3
Natural gas	0.3	0.8
Wood	-	-
Total stationary source	0.6	1.7
Total fuel combustion	20.1	57.6
Industrial processes	9.5	27.2
Agricultural burning	2.8	8.0
Solid waste disposal	2.0	5.7
Miscellaneous	0.5	1.5
Total	34.9	100.0

Organic free radicals formed from hydrocarbons in a photochemical smog (prev. Section) may initiate free radical oxidation of polymers (cf. next Section).

Some hydrocarbons emitted are more active than others, especially polycyclic aromatic hydrocarbons, which even in small amounts may photosensitize formation of singlet oxygen ($^1O_2$) by an energy transfer mechanism:

$$\text{Hydrocarbons} + h\nu \longrightarrow (\text{hydrocarbons})^{*} \qquad (29)$$

*) means excited singlet and/or triplet state of the hydrocarbon molecules.

$$(\text{hydrocarbons})^{*} + O_2 \longrightarrow \text{hydrocarbons} + {}^1O_2 \qquad (30)$$

The polycyclic aromatic hydrocarbons arise from fossil fuel combustion, but their annual average concentration is rather low, e.g. anthracene 0.05-10.0 nanograms (10^{-9}) per cubic meter.

It has been reported that aromatic polynuclear hydrocarbons are sensitizers and accelerate photooxidation of different polymers, e.g. polyolefins[41-43] and rubbers[44].

Free radical oxidation processes

Organic free radicals (R.) which are formed in photochemical smog react with oxygen and produce peroxy radicals ($RO_2\cdot$) which are rather longlived and may selectively abstract hydrogen from polymers (PH) and produce polymer radicals (P·).

$$R^{\cdot} + O_2 \longrightarrow RO_2\cdot \qquad (31)$$

$$RO_2\cdot + PH \longrightarrow ROOH + P\cdot \qquad (32)$$

Polymer alkyl radicals (P·) add molecular oxygen to produce polymer peroxy radicals (POO·) which may further abstract hydrogen from the same and/or neighbouring polymer molecule. In this way a cycle of free radical oxidations occurs and is known as "autoxodiation" process[45,46]:

$$P\cdot + O_2 \longrightarrow POO\cdot \qquad (33)$$

$$POO\cdot + PH \longrightarrow POOH + P\cdot \qquad (34)$$

Oxygen reacts fast with all types of alkyl and allyl polymer radicals formed. The oxidation may cause the polymeric materials to crosslink which causes hardening or shrinkage and may show up as stress cracking. The autoxidation may degrade the macromolecules to lower molecular weight and volatile products, causing volume decrease or stress cracking of the polymer. It may also lead to acidic products which will discolour the material or bring about further degradations.

Oxidation processes with atomic oxygen, ozone and singlet oxygen

The photochemical smog contains several oxygen species such as atomic oxygen (O), ozone (O_3) and singlet oxygen ($^1O_2$) which are highly reactive with several polymers. The concentration of O and $^1O_2$ is low and their effect is considered to be less important in the total corrosion of polymers than the effects of ozone, nitric and sulfuric acids.

Atomic oxygen (O) reacts rapidly with several polymers (PH) by abstraction of hydrogen and formation of polymer oxyradicals (PO·)[33,47,50]

$$PH + O \longrightarrow P\cdot + \cdot OH \qquad (35)$$

$$P\cdot + O_2 \longrightarrow PO_2^{\cdot} \qquad (36)$$

Highly branched polymers such as polypropylene and polymers with other links, e.g. poly(oxymethylene), are most readily attacked by atomic oxygen. Perfluorinated polymers, rubber vulcanized with sulfur, and highly aromatic polymers are the most resistant.

Ozone (O_3) plays a significant role in oxidation and degradation of several polymers such as: polyolefins, rubber, polystyrene and polyamides[32,33,45,51-58]. Ozone may abstract hydrogen and form polymer peroxyradicals (POO·):

$$PH + O_3 \longrightarrow POO\cdot + \cdot OH \qquad (37)$$

or add to double bonds with formation of ozonides, which are unstable and easily decomposed into aldehydes and acids:

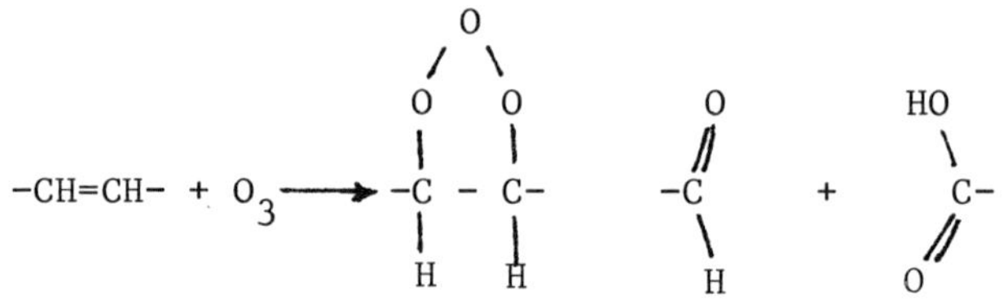

(38)

Singlet oxygen ($^1O_2$) reacts mainly with unsaturated polymers by "enetype" reactions with formation of hydroperoxy groups and shifting of the double bond[33,59,60]:

$$-CH_2-CH{=}CH- + {}^1O_2 \longrightarrow -CH{=}CH-\underset{}{\overset{OOH}{\overset{|}{C}H}}- \qquad (39)$$

As a result of these oxidation reactions polymeric materials gradually lose their useful mechanical properties. During the oxidation degradation (mainly by beta-scission processes) and crosslinking reactions occur.

Non-oxidative processes

In polluted air other degradative reactions in the absence of oxygen may also occur. The presence of water, especially acidic water, may bring about hydrolytic reactions and, if oxygen also is present, may increase the oxidation reactions. All these reactions may be going on simultaneously and in the most complicated manner. The resulting lower molecular weight units can cause changes in flow, shape, strength and other mechanical properties.

Physical changes may result from distortion of the polymeric material by sunlight due to photodegradation and photocrosslinking and can be photosensitized by the presence of impurities on the surface which efficiently absorb light. Energy transfer processes may play a considerable role.

Diffusion and migration of additives from the polymer matrix into a surface may have both accelerating or stabilizing effects. The sideeffect may be a hardening of the material from the loss of plasticizers or from the loss of other low molecular additives or even water, which can act as plasticizers. Differential expansion of organic and inorganic additives can result in cracking of the polymeric materials.

These reactions may add to the reactions caused by polluted atmosphere. They are particularly possible at temperatures near those where changes or property transitions occur (for example at softening and melting points or brittle points of the materials).

Service Life of Polymeric Materials in Polluted Atmosphere

The deterioration of plastics and other polymeric materials in polluted atmosphere takes place in a number of stages[33,61-55]:

1. Loss of gloss
2. Minute crazing
3. Severe crazing and cracking
4. Leaching
5. Loss of reinforcement from the structure
6. Gradual breakup

The timescale for these processes to occur is dependent on the type of polymer, the content of added materials, plasticizers and additives such as antioxidants, photostabilizers, thermostabilizers, etc), filling and reinforcing materials such as particles and fibers, etc.

The first observed effect is the loss of surface gloss which is an obvious sign of deterioration of a polymeric material. The gloss can be lost by abrasion from particulate matter (sometimes in rain), surface leaching and minute and severe crazing. After extended exposure times the polymeric materials may lose their mechanical properties for practical use.

Discoloration of polymeric materials is usually the result of complicated photochemical reactions and the effect of surface reaction with SO_2, NO_2, O, O_3 and singlet oxygen ($^1O_2$).

The service life of plastics and other polymeric materials is a particular problem in Los Angeles and Tokyo because of the special climatic conditions for formation of photochemical smog. This situation is by no means confined to the USA and Japan. Photochemical smog and environmental corrosion of polymers is commonly observed in some European cities, notably Madrid, Athens and London. In Stockholm where air pollution is at a low level these problems are presently not seen so serious.

Conclusions

The concentration of air pollutants is very important for the longterm properties of polymeric materials. One would expect that the problem of environmental corrosion of polymers is negligible in comparison with the physical,thermal and photochemical ageing of polymers. The tables I-IV show that annual emission of pollutants into the atmosphere is tremendous, and the effects on polymers of individual pollutants (particles, sulfur oxides, nitrogen oxides, hydrocarbons, oxygen compounds, acid rain, etc) should be taken into serious consideration and be the subject of more detailed research. The cost of environmental corrosion must include the deterioration of such diverse objects as polymeric building materials, outdoor electrical lines with polymeric insulation, rubber tires, painted surfaces, etc. The cost can only be estimated and no definite figures are available. Additional costs include the work for replacement and substitution of corroded polymeric materials, e.g. in machinery and building constructions. Observations of polymeric materials, e.g. house paint, used in heavy polluted areas demonstrate that the effects of environmental corrosion are clearly visible.

This paper does not cover all aspects of this subject. The references cited here are selected from hundreds of published reports, and they are given as examples to illustrate the importance of the problem. The need for more research is amply verified, both on the effects and their prevention.

Since 1971 our department is involved in study of photodegradation, sensitized photooxidation and photostabilization of polymers. A big effort is made in the study of singlet oxygen oxidation of several important polymers, such as polydienes, poly(vinyl chloride),polystyrene and polyesters, reactions of atomic oxygen and ozone with polydienes and photostabilization of these polymers. As a result of this work, a USA-Swedish Workshop on "Photodegradation and Photostabilization of Polymers" was organized in Stockholm in 1981[66]. At this meeting several problems included in this paper were discussed.

The workshop concluded that a broad international cooperation should be explored and current research on this problem expanded.

Areas requiring special attention in the future include improved stabilization processes for existing polymers and the synthesis of new polymeric materials which are more resistant against environmental corrosion than those presently used. The implications of the stability of natural polymers for the prediction of service life of synthetic polymers, the general effects of hydrolysis, the light-induced hydrolytic degradation are relevant research topics.

This paper includes results from our research programs on photooxidation and photo-stabilization of polymers supported by the Swedish National Board for Technical Development (STU) and a project in the NORDFORSK program on singlet oxygen reactions.

Literature Cited

1. Perkins, H.C., "Air Pollution", Mc Graw-Hill, N.Y., 1974
2. Lynn, D.A., "Air Pollution Threat and Response", Addison-Wesley, Reading Pa, 1976.
3. Holum, J.R., "Topics and Terms in Environmental Problems", Wiley, N.Y., 1978.
4. Faith, W.L. and Atkinsson, A.A., "Air Pollution", 2nd ed. Wiley, N.Y., 1972
5. Friedlander, S.K., "Smoke, Dust and Haze, Fundamental of Aerosol Behaviour", Wiley, N.Y., 1977.
6. Cavender, J.H., Kircher, D.S. and Hoffman, A.J., Nationalwide Air Pollutant Emission Trends 1940-1970, EPA Publication No. AP-115, 1973.
7. Rosato, D.V. and Schwartz, R.T., "Environmental Effects on Plastics Materials", Wiley-Interscience, N.Y., 1968,1-2.
8. Kinslow, R., Intern. Sci. Technol., 1965, 40, 38.
9. Dainton, F.S. and Ivin, K.J., Trans. Faraday Soc., 1950, 46, 374, 382.
10. Jellinek, H.H.G. and Kryman, F.J., J. Appl. Polym. Sci., 1969, 13, 2504.
11. Jellinek, H.H.G., J. Air. Poll. Contr. Ass., 1970, 20, 672.
12. Jellinek, H.H.G., in "Photochemistry of Macromolecules" (Ed. R.F. Reinish), NASA AMES Research Center, Calif. 1970, 85, 91.
13. Jellinek, H.H.G., J. Air. Pollut. Contr. Ass., 1970, 20, 672
14. Jellinek, H.H.G. and Pavlinec J., Photochem. Macromol. Proc. Symp. (Ed. R.F. Reinish), Plenum Press, N.Y., 1970, 91.
15. Hrdlovic, P., Pavlinec J. and Jellinek, H.H.G., J. Polym. Sci., A1, 1971, 9, 1235.
16. Jellinek, H.H.G., Text Res. J., 1973, 43, 557.
17. Sharovolskaya, L.N., Degtyareva A.A. and Shrubovich, V.A., Tezy Dokl. Resp. Konf. Vyskomol. Soedin., 3rd Ed. Naukova Dumka, Kiew, USSR, 1973, 87
18. Degtyareva, A.A., Kashan, A.A., Sharovolskaya L.N. and Shrubovich, V.A., Vyskomol. Soedin., 1975, A17, 2144.
19. Degtyareva, A.A., Sharovolskaya, L.N., Shrubovich V.A. and Kashan, A.A., Fiz. Khim. Polim., 1975, 15, 147.

20. Davidson, J.A. and Abrahamson, E.W., Photochem. Photobiol., 1972, 15, 403.
21. Meterological Data for Sweden, Publ. by Institute of Meteorology, Stockholm University, Stockholm, Sweden 1970-1980.
22. Jellinek, H.H.G. and Toyoshima, Y., J. Polym. Sci., 1967, A1, 5, 3214.
23. Jellinek, H.H.G. and Flajsman, F., J. Polym. Sci., 1969, A1, 7, 1153.
24. Jellinek, H.H.G., Flajsman, F. and Kryman, F.J., J. Appl. Polym. Sci., 1969, 13, 107.
25. Jellinek, H.H.G. and Flajsman, F., J. Polym. Sci., 1970, A1, 8, 711.
26. Jellinek, H.H.G. and P. Hrdlovic, J. Polym. Sci., 1971, A1, 9, 1219.
27. Jellinek, H.H.G. and Chaudhuri, A.K., J. Polym. Sci., 1972, A1, 10, 1773.
28. Jellinek, H.H.G., Techn. Report Natl. Bur. Stand (US), 1972, 357, 101.
29. Jellinek, H.H.G., Yokota, R. and Ioth, Y., Polym. J., 1973, 4, 601.
30. Jellinek, H.H.G. and Wang, T.J.Y., J. Polym. Sci., 1973, A1, 11, 3227.
31. Jellinek, H.H.G., Martin, F. and Wegener, H., J. Appl. Polym. Sci., 1974, 18, 1778.
32. Jellinek, H.H.G. Technical Report No. 782950 US NTIS, 1974.
33. Rabek, J.F., Lucki, J. and Ranby, B., Europ. Polym. J., 1979, 15, 1089.
34. Huber, H. and Joerg, F., Staub-Reinhalt. Luft, 1979, 39, 211.
35. Kambe, C.H. and Yokota, R., Polym. Eng. Sci., 1980, 20, 696.
36. Huber, H. and Joerg, F., Staub-Reinhalt. Luft., 1981, 41, 1.
37. Estefan, R.M., Gause, E.M. and Rowlands, J.R., in "The Chemistry of Air Pollution", MSS Information Co., N.Y., 1973, 55.
38. Tusday, C.S., "Chemical Reactions in Urban Atmosphere", Elsevier, N.Y., 1971.
39. "Photochemical Smog and Ozone Reactions". (Ed. R.F. Gould), ACS Ser. No. 113, Washington D.C., 1972.
40. Stephens, E.R., Calif. Air. Environmen., 1969, 1, 1.
41. Carlsson, D.J. and Wiles, D.M., J. Polym. Sci., 1973, B 11, 759.
42. Aspler, J., Carlsson, D.J. and Wiles, M., Macromolecules, 1976, 9, 691.
43. Carlsson, D.J., Garton, A. and Wiles, D.M., Macromolecules, 1976, 9, 695.
44. Rabek, J.F. and Ranby, B., Rev. Roum. Chim., 1980, 25, 1045.
45. Ranby, B., and Rabek, J.F., "Photodegradation, Photooxidation and Photostabilization of Polymers", Wiley, London, 1975.
46. McKellar, J.F. and Allen, N.S., "Photochemistry of Man-Made Polymers", Applied Science Publ., 1979.
47. Hansen, R.H., in "Interface Conversion" (Eds. P. Weiss and G.D. Cheever), Elsevier, N.Y., 1968, 287.

48. Hansen, R.H., in "Thermal Stability of Polymers" (Ed. R.T. Conley), Dekker, N.Y., 1970, 153.
49. Hansen, R.H., Pascale, J.V., DeBenedictis, T. and Rentzepis, P.M., J. Polym. Sci., 1965, A1, 3, 2205.
50. Reneker, D.H. and Bolz, L.H., J. Makromol. Chem. Sci. Chem., 1976, A10, 599.
51. Razumovski, S.D., Reutova, L.M., Niazashvili, G.A., Tutorski, I.A. and Zaikov, G.E., Dokl. Akad. Nauk SSSR, 1970, 194, 1127.
52. Razumovski, S.D., Kefeli, A.A. and Zaikov, G.E., Europ. Polym. J., 1971, 7, 275.
53. Kefeli, A.A., Razumovski, S.D., Markin, V.S. and Zaikov, G.E., Vyskomol. Soedin., 1972, A14, 2413.
54. DeVries, K.L. and Simonson, E.R., Ozone Chem. Technol., 1975, 257.
55. Lofqvist, A.R. and Haylock, J.C., Ozone Chem. Technol., 1975, 241.
56. Yamauchi, J., Ikemoto, K. and Yamaoka, A., Makromol. Chem., 1977, 178, 2483.
57. Popov, A.A., Krisyuk, B.E. and Zaikov, G.E., Vyskomol. Soedin., 1980, 22, 1366.
58. Popov, A.A., Krisuk, B.E., Blikov, N.N. and Zaikov, G.E., Europ. Polym. J., 1981, 17, 169.
59. Rabek, J.F. and Ranby, B., Photochem. Photobiol., 1978, 28, 557.
60. Golub, M.A., NASA Technical Memorandum, No 78604, 1979.
61. Jellinek, H.H.G., Kinetics Mechanisms of Polyreactions, IUPAC Int. Symposium Macromol. Chem., Plenary Lecture, 1971, 747.
62. Jellinek, H.H.G., in "Recycling and Disposal of Solid Wastes" (Ed. T.F. Yen), Ed. Ann Arbor Sci., Mich., 1974, 185.
63. Fladsman, F., Kem. Ind., 1974, 23, 96.
64. Zaikov, G.E., Khim. Zhizn., 1980, 20.
65. Seymour, R.B., "Plastics vs. Corrosives", Wiley, N.Y., 1982.
66. Ranby, B. and Vogl, O., Polymer News, August, 1982, 188.

RECEIVED April 11, 1983

18

Influence of Sulfur Dioxide on Organic Coatings

W. FUNKE and H. HAAGEN

Institut für Technische Chemie, Universität Stuttgart und Forschungsinstitut für Pigmente und Lacke, E.V., Stuttgart, Federal Republic of Germany

As far as organic coatings are concerned, SO_2 is one of the most important deteriorating factors of the environment. SO_2 may react directly with metal surfaces to form sulfates, either before or after the coating is applied. In case of steel such sulfates stimulate atmospheric corrosion. Reactions are also possible with vehicles and pigments, giving rise to degradation and discoloration. Such reactions are sometimes assisted by oxygen and sunlight. Special emphasize will be given to the permeability of organic coatings to SO_2. Though literature and data are still scanty, experimental evidence exists that SO_2-permeability is quite remarkable. Problems and methods to neutralize or capture permeating SO_2 in organic coatings by special film components will be discussed.

The increasing emission of SO_2 into the atmosphere (Table I) raises many serious problems, one of which is the influence of this pollutant on organic coatings and on their protective function[1-5]. There are various important aspects of this problem which shall be subsequently reviewed and discussed.

Corrosion Protection

It is well known that $SO_4^=$ anions stimulate corrosion of steel surfaces by preventing an in situ formation of iron oxides, which may impede the diffusion processes involved in the corrosion reactions[6]. These sulfate anions are either formed in the atmosphere by oxidation of SO^2 or by direct reaction with the steel surface in the presence of water to form so-called sulfate nests[7]. The latter transformation may take place at the unprotected metal surface or possibly, at least in principal, after SO_2 has permeated the organic film and arrived at the metal support. The diffusion of $SO_4^=$ anions through organic coatings seems

0097-6156/83/0229-0309$06.00/0
© 1983 American Chemical Society

Table I.
Characteristic SO_2 Concentration Levels

Condition	ppm (v/v)	mg/m^3	Lit.
Emission Value Longtime Exp. (TA Luft)	0.05	0.134	1
Emission Value Short Time Exp. (TA Luft)	0.15	0.4	1
MEK-Value, Exp. Time 30 min.	0.37	1.0	1
24 hrs.	0.11	0.3	1
1 yr.	0.04	0.1	1
Perception of Smell	0.5	1.34	1
MAC-Value	2.0	5	1
Smog Alarm	0.8	2.14	1
Mortal Danger	400	1068	1
Kesternich Test			
0.2 l SO_2/300 l	ca. 600	ca. 1600	2
2.0 l SO_2/300 l	ca. 6000	ca. 16000	2
Large Towns ($2x10^6$ inhab.)	0.15	0.4	3
max. value	0.33	0.9	4
Medium Towns (0.5-$2x10^6$ inhab.)	0.05	0.14	3
Smaller Towns ($0.5x10^6$ inhab.)	0.04	0.13	3
Rural Station	0.0001- −0.0075	0.001- − 0.02	5
Urban Station	0.0075- −0.015	0.02- − 0.04	5
Industrial Station	0.03	0.08	5
Industrial Area (Ruhr, West Germ.)	max. 1.28	max. 3.4	4
SO_2-conc. Stuttgart			
Summer	max. 0.015	max. 0.04	1
Winter	max. 0.05	max. 0.14	1

1 ppm=1 cm^3/m^3=2.67 mg/m^3
MEK=Maximal Emission Concentration
MAC=Maximal Allowable Concentration - Threshold Limit Value

not to play an important role. Though little information is available on $SO_4^=$ permeability, it may be assumed that it is lower than that of Cl^-[8].

SO_2 Permeability

A literature search reveals disappointingly little information on SO_2 permeabilities of organic coatings. SO_2 permeabilities of alkyd resins are treated in a series of papers[9-11]; however the published data are difficult to compare with other data because they are not presented in conventional dimensions. Only slightly more data are available on polymer films in general, part of which have been compiled (Table II). Because of differing measuring conditions and lacking informations on these conditions as well as on the materials used, even the little data on polymer films given by various authors are hard to compare[12-18]. Therefore, the only conclusions that can be drawn from this list are that SO_2 has low permeability in glassy polymers, very high permeability in rubber-like polymers, and that there is a significant decrease in SO_2 permeability with increasing polymer crystallinity.

Pressure Dependency of SO_2 Permeability

As may be seen from Table I, concentrations of SO_2 in the atmosphere are subjected to considerable variations. Moreover, SO_2 concentrations used in corrosion tests are larger by several orders of magnitude. Therefore, it becomes highly questionable whether data obtained at high SO_2 concentration levels may be also applied to practical exposure conditions. SO_2 permeabilities of a series of polymer films like polyethylene, polycarbonate, polyamide[12] as well as polyacrylate and cellulose triacetate[19] have been shown to depend on SO_2 pressure, especially at higher SO_2 concentration levels. In the case of organic coatings such data are not available. It is therefore recommended to study SO_2 permeability at low SO_2 concentrations comparable to practical exposure conditions.

Comparison of SO_2 Permeability with Permeabilities of Other Gases

Contrary to other atmospheric gases like N_2, O_2 or CO_2, the permeability of polymer films for SO_2 is very high (Table III). Obviously the molecular size of SO_2 is not the dominating factor for its permeation rate. As the permeability coefficient $\underline{P}$ is defined by the product of the diffusion coefficient, D, and the solubility, S, the SO_2 solubility of the polymer films plays an important role. In fact the few data published on solubility of SO_2 in polymers corroborate this expectation. Equilibrium solubility of SO_2 in polyacrylate was found to be as high as 21.5% by weight at 760 mmHg[19]. In case of a bisphenol A polycarbonate,

Table II.
SO_2 Permeabilities of Polymer Films

Polymer Film	Temp. °C	Thickness m	PSO_4	Lit.
Polyethylene (low density)	25	230	20.9	12
Polyethylene (high density)	25		5.68	13
Polyamide (Nylon 11)	a) 25	61.4	6.58	12
	b) 25	41	2.16	13
Polycarbonate (Lexon)	a) 25	131	22.4	
	b) 25	25	21.0	13
Vinylidene Chloride-Vinylchloride Copolymer	25		0.201	12
Polyethylene Terephthalate	a) 22		5.27	14
	b) 25	13	0.201	13
Polymethylmethacrylate	22		0.132	14
Polyvinylchloride (rigid)	25	145	0.116	13
Teflon	25		4.2	15
Polystyrene	25	38	22	13
Cellulose (cellophane)	22		52.7	16
Cellulose Nitrate	25		176	17
Vulcan Rubber	22		1450	14
Silicone Rubber	22		11800	14

Table III.
Comparative SO_2, O_2, N_2 and CO_2 Permeabilities of Polymer Films

Polymer Film	PSO_2/PO_2	Lit.
Polyethylene (230 m)	6.8	12
Polyamide (61 m)	47	12
Polycarbonate (131 m)	15	12
Vinylidene Chloride - Vinyl Chloride Copolymer	65	12
Polyacrylate (22.6 m)	515[1]	19
Polyacrylate (22.6 m)	19.2[2]	19

1. PSO_2/PN_2
2. PSO_2/PCO_2

13.5% by weight was observed[20] and explained by electron-donor-acceptor complexes. Data of weight increase of non-pigmented organic coatings on 24 hrs exposure to SO_2 at 760 Hg support these results (Table IV). Quite obviously many polymer films have a high affinity to SO_2. SO_2 absorption by polybutylmethacrylate films was strongly influenced by the solvent used in film formation (Table V). Probably in this case the total absorption is composed of two concurrent processes: cluster formation or filling of holes which obeys a Langmuir isotherm and ordinary dissolution which obeys the Henry Law. As has been shown earlier[21], film formation may be accompanied by phase separation leading to incoherent microscopic or submicroscopic solvent inclusions, which may provide the sites for the Langmuir absorption process. Again the questions remain whether both processes also occur at practical SO_2 concentration levels and how much SO_2 is absorbed there. Moreover, a gradual release of retained solvent and stress-relaxation assisted by SO_2 absorption may give rise to time dependency in absorption measurements.

Table IV.
Weight Increase of Polymer Films After 24 Hours Exposure to Pure SO_2 at Atmospheric Pressure[26]

Polymer	Weight Increase %
Cellulose Nitrate	3
PVC-Copolymer	5
Alkyd/Melamine-Resin	14
Epoxy-Resin	11
Polyurethane/Acrylate	6

Table V.
Influence of Solvents Used in Film Formation on SO_2 - Absorption After 24 Hours of Exposure[26]

PBMA Prepared from Solution With	Weight Increase %
Mineral Spirits	15
Toluene	16
MIBK	11
Ethylacetate	6
Isopropanol	12

Effect of Humidity and Pigmentation on SO_2 Permeability

Increasing humidity also increases SO_2 permeability. However, no significant effects can be observed if water absorption of polymer films is low. As most organic coatings absorb only small amounts of water, normal variations in humidity should not significantly affect SO_2 permeation. As with other fillers, pigmentation significantly reduces SO_2 permeability[22] by the same reason as does polymer crystallinity.

Reaction of the Film Forming Polymers and Pigments with SO_2

As a result of exposure to SO_2, embrittlement and discoloration of polymer films have been observed. For chemical reactions to take place, usually the presence of water, oxygen and sometimes UV-radiation is required. With hydrocarbons and UV-radiation, sulfinic acids are formed[23]. No reaction takes place with polyethylene in absence of light. Polyethylene and polypropylene crosslink on exposure to light and SO_2. Besides a direct reaction with the polymer, a photochemical reaction of activated SO_2 with oxygen may lead to formation of ozone which then may attack polymer films by oxidative degradation. A series of inorganic pigments like chrome yellow, chrome-molybdate and ZnO are attacked by SO_2 with formation of Cr (III) sulfate, lead sulfate, and $ZnSO_4$[24], respectively. Heavy degradation and discoloration were observed on exposure of coated aluminum wall panels to SO_2 and UV-light[25]. Active corrosion protective pigments like red lead may neutralize the corrosion-stimulating action of SO_2 by formation of slightly soluble $PbSO_4$. As these pigments are increasingly prohibited because of toxic effects and environmental protection, it is imperative to incorporate other SO_2 scavenging components in organic coatings to increase their protection efficiency.

Literature Cited

1) J. Baumuller and U. Hoffmann, Bauphysik, 1, 22 (1982).
2) German Standardization DIN 50 018.
3) H. C. Wohlers and G. B. Bell, "Literature review of metropolitan air pollutant concentrations", Stanford Res. Inst. Report, 1958.
4) J. Ruf, "Korrosion", Schutz durch Lacke und Pigmente, Verlag W. A. Colomb, Stuttgart, 1972.
5) L. G. Johanson and N. G. Vannerberg, Werkstoffe und Korrosion, 32, 265 (1981).
6) K. Barton, S. Bartonova and E. Beranek, Werkstoffe und Korrosion, 25, 659 (1974).
7) H. Schwarz, Werkstoffe und Korrosion, 16, 93 (1965).
8) A. L. Glass and J. Smith, J. Paint Technology, 39, 490 (1967).

9) M. Svoboda, H. Klicova and B. Knapek, Farbe und Lack, 74, 659 (1968).
10) M. Svoboda, H. Klicova and B. Knapek, J. Oil Chem. Assoc., 52, 677 (1969).
11) M. Svoboda, H. Klicova and B. Knapek, Farbe und Lack, 75, 121 (1969).
12) E. G. Davis and M. L. Rooney, Kolloid Ztschr., Ztschr. f. Polymere 249, 1043 (1971).
13) E. G. Davis, M. L. Rooney and P. L. Larkins, J. Appl. Polym. Sci., 19, 1829 (1975).
14) M. Benarie and Bui-the-Chuong, Atmos. Environ., 3, 574 (1969).
15) B. E. Saltzman, C. R. Feldmann and A. E. O'Keefe, Environ. Sci. Technol., 5, 1121 (1974).
16) J. Hanonsek and L. Herynk, Chem. Listy, 56, 376 (1962).
17) P. Y. Hsieh, J. Appl. Polym. Sci., 7, 1743 (1963).
18) R. M. Felder, R. D. Spence and J. K. Ferrell, J. Chem. Eng. Data, 20, 235 (1975).
19) D. L. Kuehne and S. K. Friedlander, Ind. Eng. Chem. Process Dev., 19, 609 (1980).
20) E. G. Davis and M. L. Rooney, J. Polym. Sci., 10, 2325 (1972).
21) W. Funke, J. Oil Chem. Assoc., 59, 398 (1976).
22) M. Svoboda, G. Klicova and B. Knapek, Zashchita Metallov, 9, 301 (1973).
23) H. H. G. Jellinek, F. Flajsman and F. J. Kryman, J. Appl. Polym. Sci., 13, 107 (1969).
24) S. Torlaski, L. Caggiati and G. Zorzella, "Fatipec Conference Florence 1972", Conference book p. 207, 1972.
25) H. Haagen - Unpublished results.
26) H. Haagen - Farbe und Lack, in print.

RECEIVED April 4, 1983

19

Photochemical Degradation and Biological Defacement of Polymers

P. D. GABRIELE, J. R. GEIB, R. M. IANNUCCI, and W. J. REID

Ciba-Geigy Corporation, Additives Department, Ardsley, NY 10502

Engineering plastics and organic coatings are being exposed to increasingly more hostile environments as they replace and protect more wood and metal products, especially in exterior applications. When exposed outdoors these plastics must withstand the concerted effects of moisture, oxygen, heat, ultraviolet light, and micro-organisms in order to perform their designed function. The combined effects of the first four factors, i.e., moisture, oxygen, heat and UV light, can result in photooxidation in the polymer surface. Eventually, erosion and microfracture of the polymer surface ensues, thus creating a microenvironment which is conducive to moisture and "dirt" accumulation. Also, the photooxidation leads to high concentrations of a variety of carbonyl groups which cause the surface to be hydrophilic. This hydrophilicity allows the polymer surface to absorb moisture causing it to swell and result in additional stress cracking. The surface microenvironment provides an ideal atmosphere for mildew growth, i.e., free moisture, carbon and nutrient sources which accumulate during the weathering process. The polymers and coatings which we have examined have had to undergo significant surface deterioration prior to the time that mildew could undergo active growth on the surface. By proper stabilization of the polymers and coatings with benzotriazole UV absorbers, hindered amine light stabilizers (HALS), or combination of the two stabilizers, the surfaces of the polymers and coatings could be maintained and thus active mildew growth prevented.

0097–6156/83/0229–0317$06.00/0
© 1983 American Chemical Society

Engineering plastics and organic coatings are being exposed to increasingly more hostile environments as they replace and protect more wood and metal products, especially in exterior applications. When exposed outdoors these materials must withstand the concerted effects of moisture, oxygen, heat, ultraviolet light, and micro-organisms in order to perform their designed function. Photodegradation of plastic materials that have been exposed to outdoor conditions has been studied extensively (1). Biological degradation of plastic materials has also been widely studied (2,3,4). Recently, we reported the results of numerous studies of different plastic materials that had been exposed outdoors in Florida (5). The purpose of these studies had been to show the effectiveness of hindered amine light stabilizers (HALS), typically bis[2,2,6,6-tetramethyl-4-piperidinyl]sebacate (LS I), and benzotriazole UV absorbers, typically 2(2'-hydroxy-5'-methyl-phenyl)benzotriazole (LS II), used alone or in combination with each other in preventing the photodegradation of plastic materials. We had shown that the optimally stabilized samples of a "weatherable" styrenic terpolymer (an ABS-type plastic that contained a saturated rubber in place of the butadiene) stabilized with a combination of 0.5% LS I and 0.5% LS II had been able to maintain its color, appearance, surface integrity, and physical properties through four years of outdoor weathering in Florida. The control with no light stabilizers was pitted and heavily encrusted with mildew. Other samples in this study which contained only 0.5% HALS or only 0.5% benzotriazole, were also heavily encrusted with mildew at this same time period, 4 years, thus indicating that neither component of the blend was functioning as a fungicide. Also, mildew was actively growing on the rough cut sides of the optimally light stabilized samples, again indicating that neither of the light stabilizers alone or in combination with each other was acting as a fungicide. (Figure 1). Samples of polypropylene stabilized with just LS I had also been able to maintain their color, appearance, surface integrity, and physical properties through 2.5 years of outdoor weathering in Florida. Moreover, thermoplastic polyurethane materials stabilized with just LS I also had excellent performance through six months outdoor weathering in Florida. All of the above plastic materials that contained no light stabilizers, low amounts of the preferred light stabilizers, or other types of light stabilizers had extensive surface erosion and heavy infestations of micro-organisms, most notably mildew (*Aureobasidium pullulans*), after the same periods of outdoor exposure.

Mildew defacement of organic coatings and plastics has long been a major problem of the organic coatings and plastics industries. *Aureobasidium pullulans* is the major causitive organism

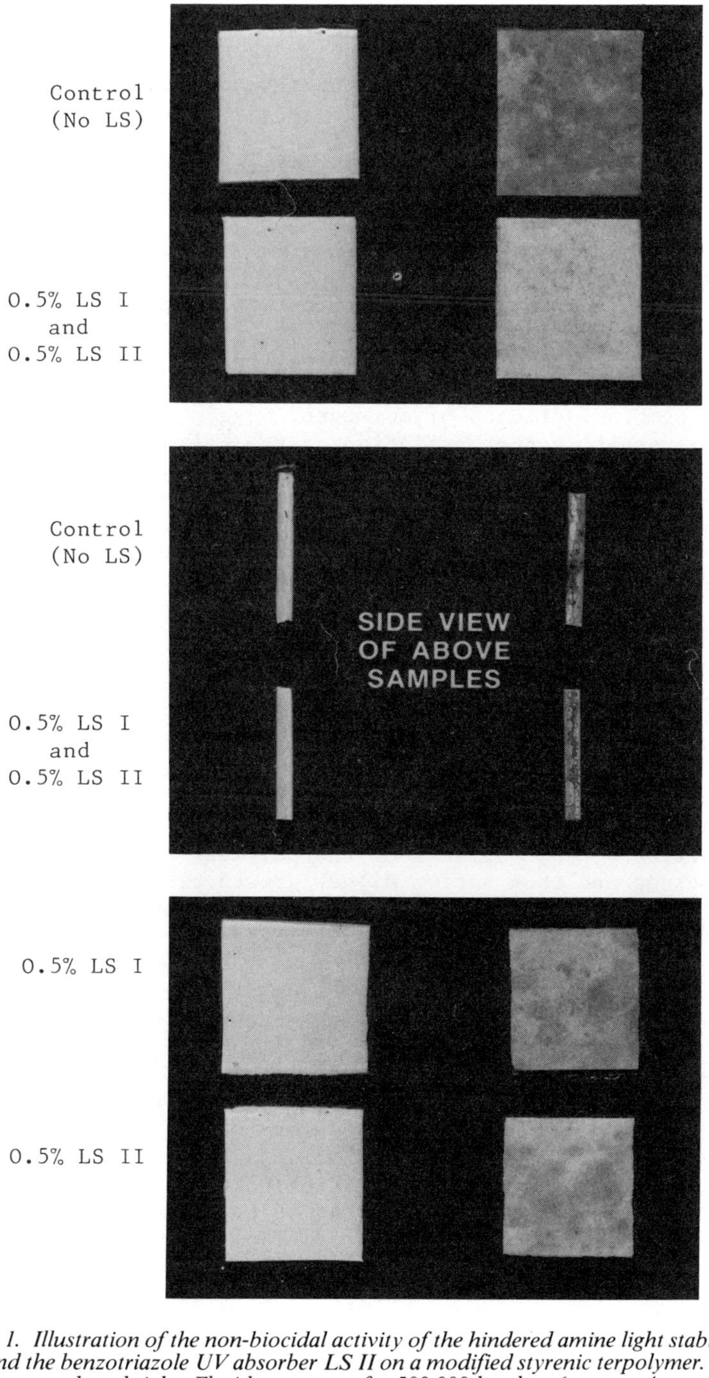

Figure 1. Illustration of the non-biocidal activity of the hindered amine light stabilizer LS I and the benzotriazole UV absorber LS II on a modified styrenic terpolymer. Key: left, unaged; and right, Florida exposure for 500,000 langleys (approx. 4 years).

responsible for this defacement (5-8). It was the primary organism identified as being responsible for the biological defacement in the three polymer systems without light stabilizers that we had previously examined. Until now, the approach to solve this problem has been to incorporate a fungicide into a coating formulation or compound a fungicide into a plastic. Although this technique had success for short service time periods, the fungicide may be vaporized or leached from the coatings and plastics over long time periods. In addition, the biological activity of most non-mercurial organic and organometallic fungicides are significantly destroyed by photo-oxidation from the ultraviolet light present in terrestrial solar radiation. Once the concentration of the fungicide drops below critical levels, mildew can attack the coating or plastic surface (6).

The combined effects of moisture, oxygen, heat, and UV light can result in photooxidation of the polymer surface. Eventually, erosion and microfracture of the polymer ensues, thus creating a microenvironment which is conducive to moisture and "dirt" accumulation. Also, the photooxidation leads to high concentrations of a variety of carbonyl groups which cause the surface to be very hydrophilic. This hydrophilicity allows the polymer surface to absorb moisture causing it to swell and result in additional stress cracking. The surface microenvironment provides an ideal atmosphere for mildew growth, i.e. free moisture, carbon, and nutrient sources which accumulate during the weathering process. The polymers and coatings which we have examined have undergone significant surface deterioration prior to the time of mildew appearance. By proper stabilization of the polymers and coatings with benzotriazole UV absorbers, hindered amine light stabilizers (HALS), or combination of the two stabilizers, the surfaces of the polymers and coatings could be maintained and thus active mildew growth prevented. In addition, by preventing the photooxidation of the surface of the plastic by the addition of the above light stabilizers, both the appearance and the physical properties of the polymers are preserved for significantly longer periods of time than if they were unstabilized or stabilized with other types of stabilizers.

In the current study, samples of impact polystyrene that contained a combination of light stabilizers LS I and LS II and samples that contained no light stabilizers were weathered outdoors in Florida and were monitored for changes in physical appearance and for changes in the chemical structure of the surface by use of a multiple internal reflectance IR spectrophotometer. Also, samples of a thermoplastic polyurethane that was stabilized with just LS I and samples that contained no light stabilizers were weathered and monitored in the same way as the impact polystyrene. We hope to establish at what point in the photodegradation of a polymer it becomes susceptible to attack by microorganisms and establish what effect the light stabilizers have in preventing or delaying the photodegradation and subsequent attack by the microorganisms.

EXPERIMENTAL

1. Preparation and Light Exposure of Polymer Samples.

(a) Impact Polystyrene: Light stabilizers were compounded into commercial impact polystyrene (IPS) by a commercial processor. The samples were injection molded into 1/8 inch thick discs and were subsequently weathered at 45° facing south in Florida. Sample 1 was impact polystyrene which contained no light stabilizers and Sample 2 was impact polystyrene which contained 0.5% LS I plus 0.5% LS II.

(b) Thermoplastic Urethane Films: Commercial thermoplastic polyurethane (Estane 5707 from B. F. Goodrich) was dissolved in a 1:1 (w/w) mixture of DMF/toluene to form a 20% solution. Light stabilizers were separately dissolved in a minimum of the 1:1 solvent mixture and then were added to the resin solution. Drawdowns of the resin solutions were prepared to obtain a resultant 1.5 mil thick film after drying. The drawdowns were flashed for five minutes at room temperature and then were forced-air dried for 15 minutes at 75°C. Subsequently, the films were exposed at 45° under glass facing south in Florida. Sample 3 was a thermoplastic polyurethane which contained no light stabilizers and Sample 4 was a thermoplastic polyurethane which contained 0.5% LS I.

2. Scanning Electron Microscope (SEM).

Each sample was cemented to a specimen holder and then was coated with approximately 200 Å of aluminum in an EFFA Rotary Vacuum Evaporator. The photomicrographs were obtained from a Cambridge Model S180 Scanning Electron Microscope.

3. Multiple Internal Reflectance (MIR) Infrared Spectroscopy.

All samples were run on a Perkin Elmer 281 with data station using the multiple internal reflectance attachment (part #186-0382). A 45° reflection angle was used for all samples.

RESULTS

The samples of impact polystyrene and a thermoplastic polyurethane were weathered in Florida, and then examined with a scanning electron microscope and with a multiple internal reflectance (MIR) spectrophotometer with the following results.

Within six months of outdoor exposure in Florida, the samples of both IPS and polyurethane that contained no light stabilizers had active growths of mildew present. No mildew was evident on the stabilized samples after the same period of exposure. (Figures 2, 3, 4)

Photos 1a and 2a are the photomicrographs of the unexposed impact polystyrene and illustrate the initial integrity of the polymer surface.

Spectra 1a is a MIR spectra of unexposed neat IPS. Spectra 2a is a MIR spectra of the unexposed IPS that contained light stabilizers. Note the intense aromatic ring breathing modes between approximately 1600 cm^{-1} and 1400 cm^{-1} (9).

Photos 1b and 2b are photomicrographs of IPS samples that had been weathered outdoors in Florida for three months.

Spectra 1b illustrates the buildup of carbonyl groups at about 1700 cm^{-1} after three months outdoor exposure of the IPS. Note the degree of band broadening in the 1600 cm^{-1} region which is probably due to aryl carbonyl formation.

Photo 1c is a photomicrograph of IPS that contained no light stabilizers which was exposed outdoors in Florida for six months. Note the total loss of surface integrity including deep surface crazes. This photo also shows the mildew that was visually present along with their typical hyphal structures.

Photo 2c is a photomicrograph of IPS that contained the combination of light stabilizers (LS I and LS II) which was exposed outdoors in Florida for six months. Note the minimal amount of surface deterioration. Some surface dirt is present in the photo.

Spectra 1c corresponds to Photo 1c. Note the intense development of a carbonyl band around 1620 cm^{-1}. After six months outdoor weathering the peaks related to the aromatic skeletal ring breathing mode have been masked.

Spectra 2c corresponds to Photo 2c. A significant carbonyl peak has formed, but the aromatic skeletal ring breathing mode peaks are still quite apparent.

Photos 3a and 4a are the photomicrographs of the unexposed thermoplastic polyurethanes and illustrate the initial integrity of the polymer surface. Spectra 3A (neat, no light stabilizer) and 4a (0.5% LS I) represent the MIR spectra of the unexposed polyester urethane.

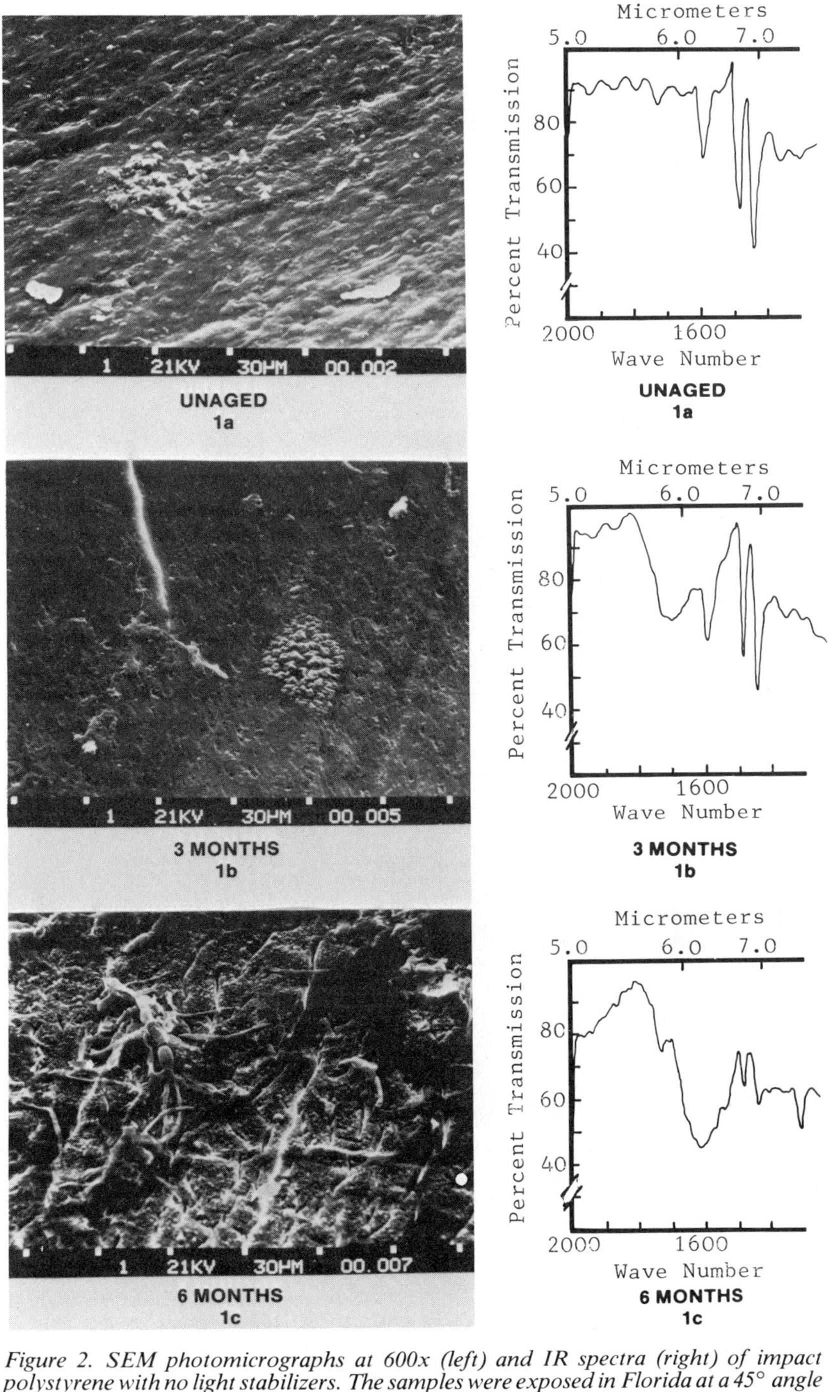

Figure 2. SEM photomicrographs at 600x (left) and IR spectra (right) of impact polystyrene with no light stabilizers. The samples were exposed in Florida at a 45° angle facing south. SEM shows the presence of mildew.

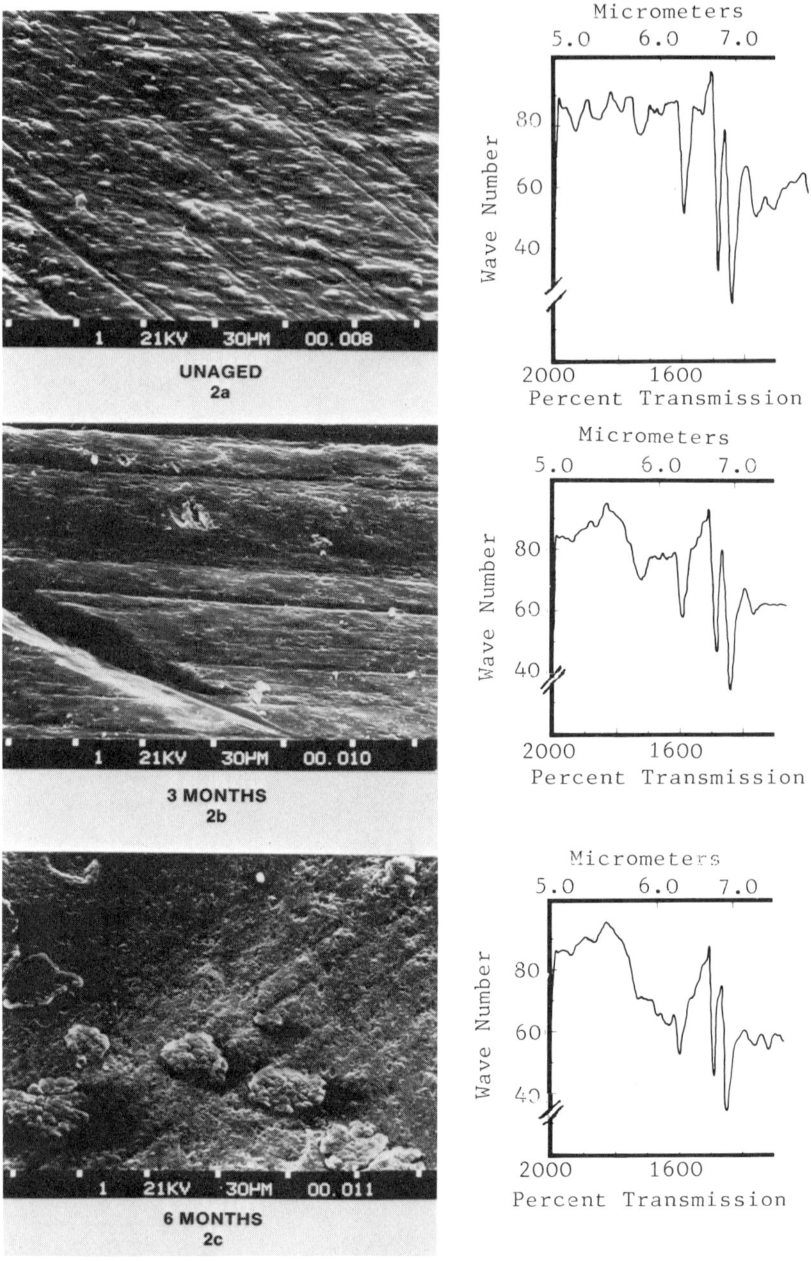

Figure 3. SEM micrograph at 600x (left) and IR spectra (right) of impact polystyrene with 0.5% LS I (HALS) and 0.5% LS II (benzotriazole UV absorber). The samples were exposed in Florida at a 45° angle facing south. The 6-month SEM shows the presence of some surface dirt.

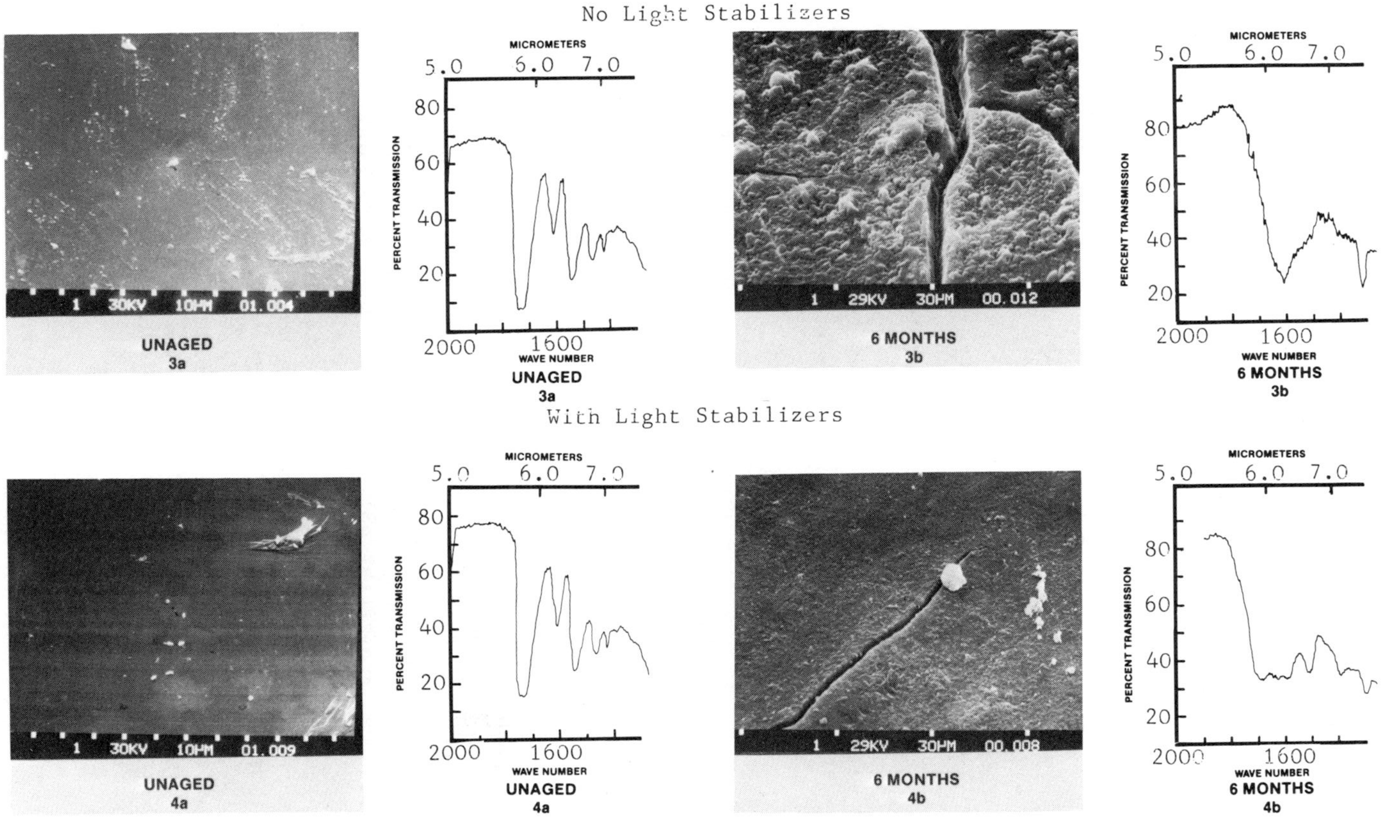

Figure 4. SEM micrograph at 1001x and IR spectra of thermoplastic polyester urethane exposed in Florida at a 45° angle under glass facing south. The light-stabilized sample contained 0.5% LS I (HALS).

Photo 3b is a photomicrograph of the polyurethane that contained no light stabilizers which was exposed outdoors in Florida for six months. Note the deep cracks and extensive surface erosion as well as mildew near the cracks.

Photo 4b is a photomicrograph of the polyurethane that contained 0.5% of LS I which was exposed outdoors in Florida for six months. The photo shows what may be the onset of surface cracking but no overall surface erosion is evident. Additionally, mildew is virtually non-existent.

Spectra 3b corresponds to Photo 3b. Note the large carbonyl band around 1700 cm^{-1} and the total loss of the 1525 cm^{-1} band which represent C-N vibration.

Spectra 4b corresponds to Photo 4b. Even though new carbonyl development has occurred, the original 1700 cm^{-1} and 1525 cm^{-1} bands are still quite apparent.

After 9 months of outdoor exposure in Florida (Figure 5), and even after 12 months (Figure 6), the SEM photomicrographs of the IPS (impact polystyrene) show minimal surface deterioration for the samples that contain 0.5% LS I and 0.5% LS II, while the samples that contain no light stabilizers show extreme surface erosion and the presence of mildew.

DISCUSSION

This study indicates that chemical changes develop during polymer surface photo-oxidation and apparently precede biological defacement. In contrast, optimum light stabilization of polymeric materials can reasonably delay premature biological defacement. The controlling factor in the defacement process is the disruption of surface integrity and the development of microfractures and cracks which predisposes accumulations of moisture and organic debris.

We have demonstrated that in styrenic polymers and in a urethane coating that surface carbonyl group development is a major chemical change following prolonged sunlight exposure.

The chemical changes which accompany photo-oxidation apparently predisposes the polymer to attack by micro-organisms. Previously, we reported that the physical changes which accompany severe photo-oxidation provide the perfect environment for debris accumulation to support the active growth of microorganisms (5).

The UV portion of terrestial sunlight causes photooxidation of the polymeric material and yields high carbonyl content, high polarity, increased hydrophilicity, and decreased molecular weight (1). These changes favor development of microorganisms. Numerous laboratory studies have demonstrated

No Light Stabilizers
(300x, top; 1000x, bottom)

With Light Stabilizers
(300x, top; 1000x, bottom)

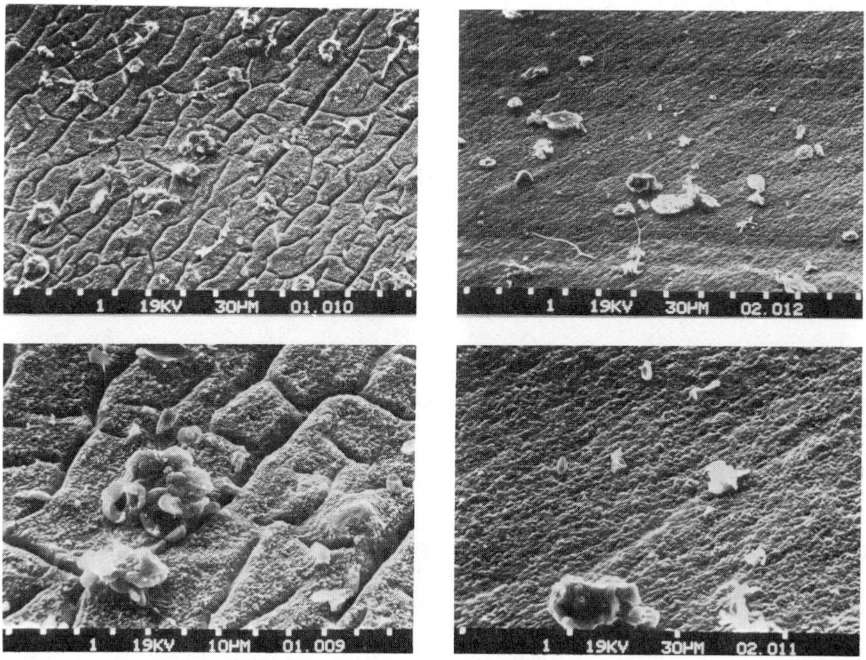

Figure 5. SEM photomicrographs of impact polystyrene exposed in Florida at a 45° angle facing south for 9 months. The light-stabilized samples contained 0.5% LS I (HALS) and 0.5% LS II (benzotriazole UV absorber).

No Light Stabilizers (300x, top; 1000x, bottom) | With Light Stabilizers (300x, top; 1000x, bottom)

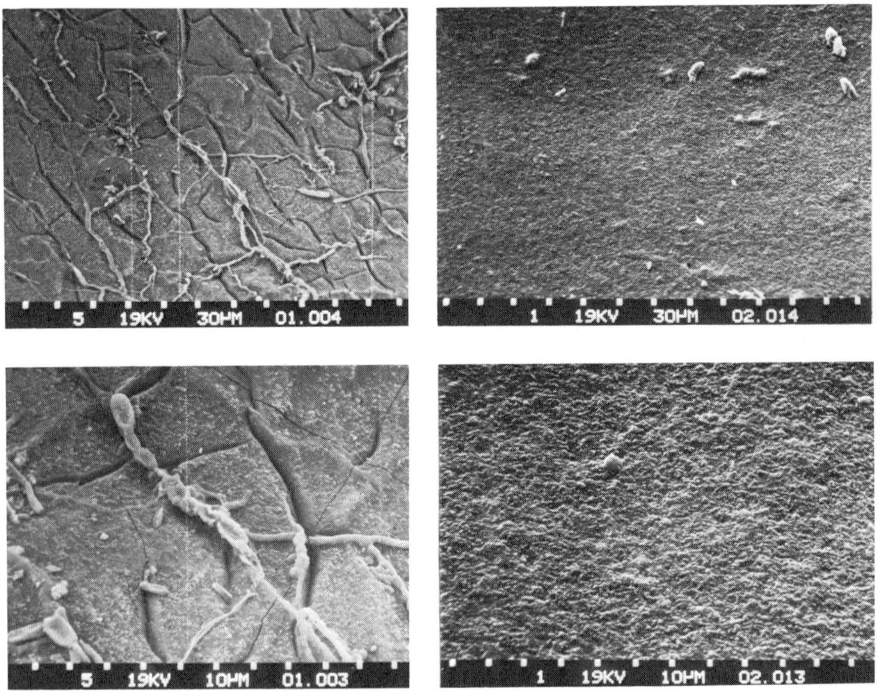

Figure 6. SEM photomicrographs of impact polystyrene exposed in Florida at a 45° angle facing south for 12 months. The photos of the non-light-stabilized samples show the presence of mildew. The light-stabilized samples contained 0.5% LS I (HALS) and 0.5% LS II (benzotriazole UV absorber).

the ability of microbes to inhabit artificially photo-oxidized materials (2,3,4). It is reasonable to assume there is a dynamic photooxidative/ bio-oxidative interrelationship involved in defacement. Therefore, defacement may be part of a natural sequence of events in nature to reduce the amount of total organic material in the biosphere.

The chemical and physical observations here indicate that biological defacement of polymeric materials is a process precipitated by surface changes. This study suggests that polymers in exterior hostile environments can have significantly delayed surface erosion by the incorporation of light stabilizers. Optimal stabilization of the polymer surface with benzotriazoles and/or hindered amine light stabilizers provides the maintenance of surface integrity and thus increases the material's functional lifetime.

The concept of light stabilization preventing biological defacement is unique and differs dramatically from the classic approach of dispersion of toxicants throughout a polymer to provide a biologically-free surface. Incorporation of a toxicant into a polymer may provide a biologically-free surface but it does not address the inevitable process of photo-oxidation.

CONCLUSION

By use of this technology, polymeric materials can be kept aesthetically attractive in hostile environments without excessive usage of toxicants. A major advantage of this is in the reduction of formulation complexity. Plastics and coatings manufacturers are limited as to the types and amounts of fungicides they can use on a cost performance basis. The current approach to a longer service life is to increase the amount of light screening pigments like ZnO and TiO_2. However, fungicide behavior can be dramatically affected by the formulation in which it is prepared. As formulations become complex, behavior of the entire formulation may be affected due to ingredient incompatibility with an obvious change in performance. The use of light stabilizers, such as benzotriazoles or hindered amines (HALS) alone or in combination with each other, as we have shown, provides an attractive alternative to the use of fungicides.

Acknowledgment

We wish to acknowledge the excellent electron microscope work performed for us by Irving Nusynowitz.

Literature Cited

(1) Ranby, B. and Rabek, J. F., "Photodegradation; Photooxidation and Photostabilization of Polymers," John Wiley and Sons, Inc., New York, 1975.
(2) Walters, A. H. and Elphick, J. S. "Biodeterioration of Materials," Elsevier Publishing Company Ltd., London, Sept., 1968.
(3) Walters, A. H. and Hueck-Van der Plas, "Biodegradation of Materials," Vol. 2, John Wiley and Sons, Inc., New York, 1972.
(4) Sharpley J. M. and Kaplan, A.M., "Proceedings of the Third International Biodegradation Symposium," Aug., 1975.
(5) Gabriele, P. D., Geib, J. R., Puglisi, J. S., and Reid, W. J., " ACS Organic Coatings and Applied Polymer Science Proceedings," 46, 538-543, March, 1982.
(6) Brand, B. G. and Kemp, H. F., "Mildew Defacement of Organic Coatings," Federation of Societies for Coating Technology, Philadelphia, PA, 1978.
(7) Schmitt, J. A., "Growth Patterns of Paint Mildew and Aureobasidium Pullulans", Journal of Coatings Technology, 50, No. 639, 51-53 (1978).
(8) Ludwig, L. E., "Formulation of Mildew Resistant Coatings", Journal of Paint Technology, 46, No. 594, 31 (1974).
(9) Bellamy, L. J., "The IR Spectra of Complex Molecules," Chapman and Hall, London, 1975.

RECEIVED March 3, 1983

INDEX

D

R

S

T

Production by Anne Riesberg
Indexing by Florence Edwards
Jacket design by Anne Bigler

Elements typeset by Hot Type Ltd., Washington, DC
Printed and bound by Maple Press Co., York, PA